By Amy Stewart

The Tree Collectors: Tales of Arboreal Obsession

From the Ground Up: The Story of a First Garden

The Earth Moved: On the Remarkable Achievements of Earthworms

Flower Confidential: The Good, the Bad, and the Beautiful in the Business of Flowers

Wicked Plants: The Weed That Killed Lincoln's Mother & Other Botanical Atrocities

Wicked Bugs: The Louse That Conquered Napoleon's Army & Other Diabolical Insects

The Drunken Botanist: The Plants That Create the World's Great Drinks

Girl Waits with Gun

Lady Cop Makes Trouble

Miss Kopp's Midnight Confessions

Miss Kopp Just Won't Quit

Kopp Sisters on the March

Dear Miss Kopp

Miss Kopp Investigates

THE TREE COLLECTORS

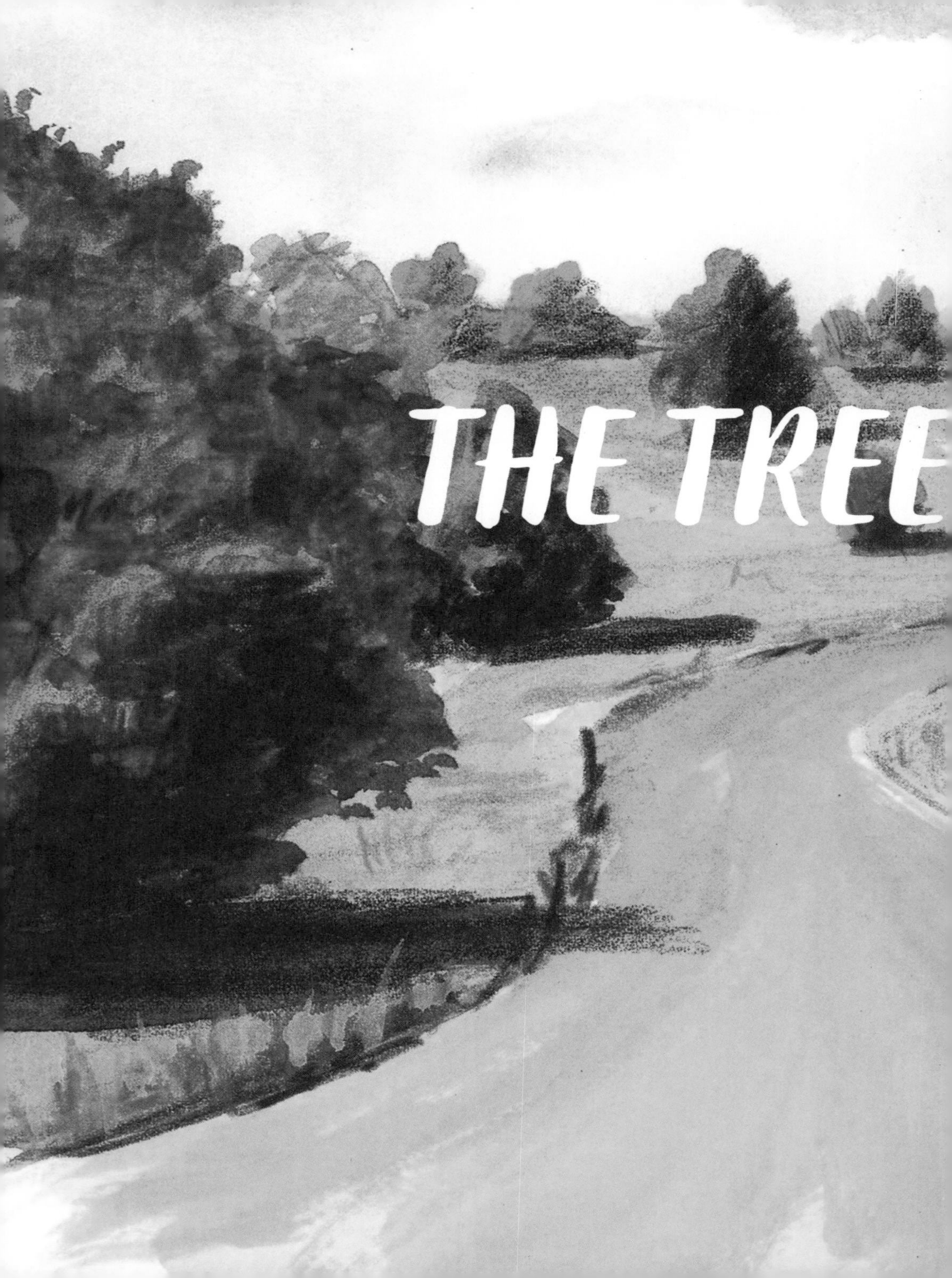
THE TREE

Tales of Arboreal Obsession

COLLECTORS

Written and Illustrated by

AMY STEWART

RANDOM HOUSE · NEW YORK

Published in the United States by Random House, an imprint and division of Penguin Random House LLC, New York.

RANDOM HOUSE and the HOUSE colophon are registered trademarks of Penguin Random House LLC.

Hardback ISBN 9780593446850
Ebook ISBN 9780593446867

Printed in Malaysia on acid-free paper

randomhousebooks.com

9 8 7 6 5 4 3 2 1

FIRST EDITION

Book design by Simon M. Sullivan

TO PSB

On the last day of the world
I would want to plant a tree.

—W. S. MERWIN

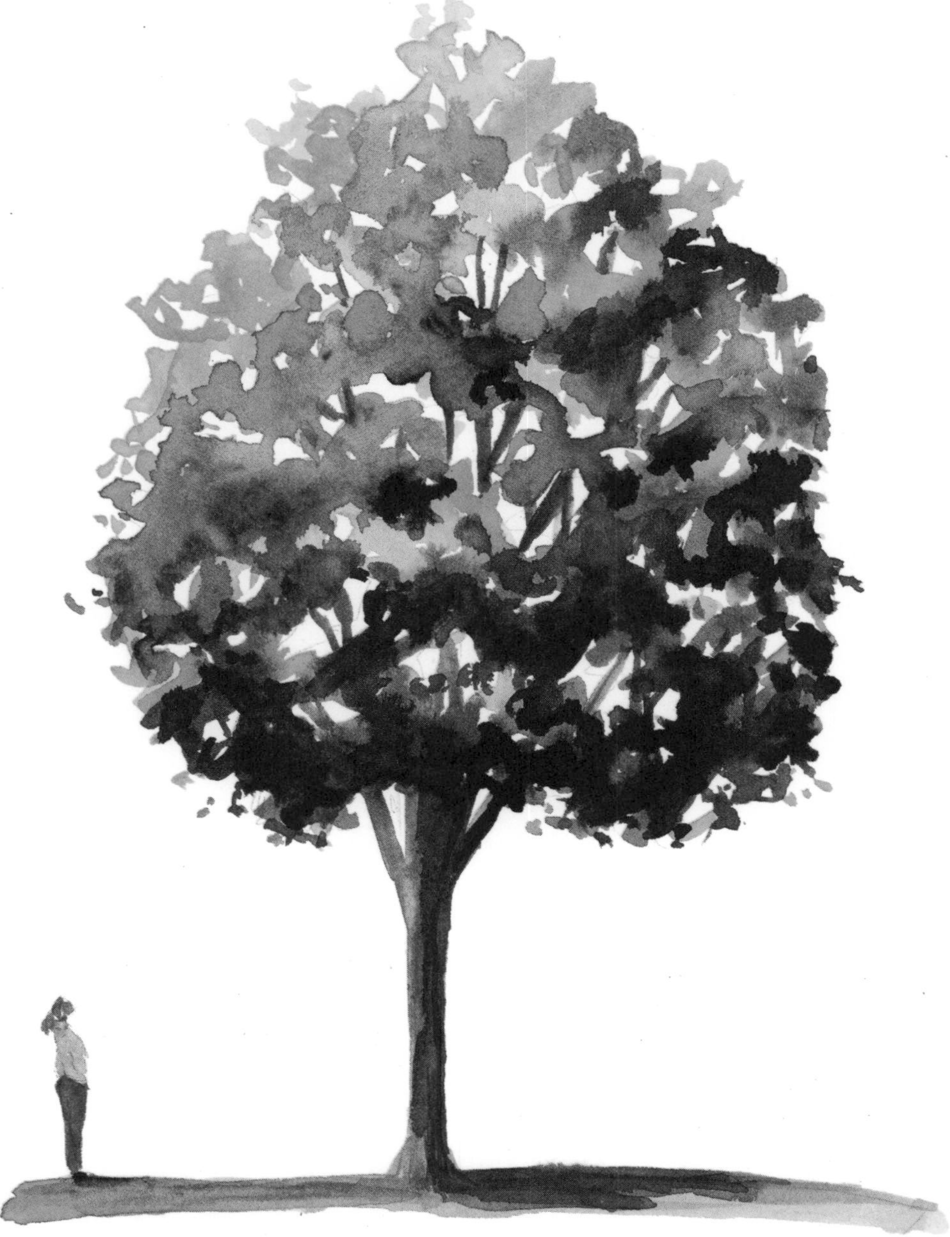

CONTENTS

HEALERS

ECOLOGISTS

ARTISTS

CURATORS

EDUCATORS

COMMUNITY BUILDERS

ENTHUSIASTS

SEEKERS

PRESERVATIONISTS

INTRODUCTION

WHAT POSSESSES SOMEONE TO POSSESS A TREE?

Ten years ago, a man named Len Eiserer introduced himself to me as a tree collector. He lived in Pennsylvania, where he had some acreage. He planted every kind of tree that struck his fancy, cramming them together as closely as he reasonably could. He told me that he had planted 150 different species and cultivars in rows, the way a book collector would line up books on a shelf.

I remember thinking that trees were an odd thing to collect. They're large and difficult to move. Collectors' tastes can change over the years: if you're into toy cars or antique buttons, you'll grow tired of the more common ones and want to sell them off in favor of something rare and exquisite. But how do you do that when your collection consists of enormous living, breathing organisms?

Then I met Dave Adams, from Boise, Idaho, whose collection of tropical trees couldn't survive the winter. At the first sign of snow, his guest room and garage became a sort of hotel for trees. He was completely out of space, but he still relied on a tropical tree dealer in San Diego to feed his habit.

Now I knew two tree collectors.

In 2019, the poet W. S. Merwin died, and his obituaries described the extraordinary collection of palms he'd assembled on his land in Hawaii.

That was the final straw. Tree collectors were (pardon the expression) coming out of the woodwork. If I'd discovered three without trying, how many more would I find if I went looking?

Plenty, as it turned out. Just like coin collectors or stamp collectors, tree collectors like to hang out with one another. They form clubs, they hold seminars and swap meets, they share pictures in online forums, and they go on field trips together. Whether you're obsessed with oak trees, maples, conifers, or magnolias, there's a tree society for you.

I started hanging out in these groups and asking people if they would talk to me about their collections. At first, nobody could believe that I wanted to hear about them. Their friends didn't understand their compulsions. Their spouses had long since grown tired of tree talk. This is the reason they form societies in the first place: as magnolia collector Beth Edward said, "No one else that I know in my career, or even in my family, really shares this interest of mine. I'm the only one. When I'm with the people in the Magnolia Society, I'm just another person. I'm normal."

When I started talking to tree collectors, I thought we'd have geeky horticultural conversations about rare cultivars and obscure subspecies. I expected to hear about trips to remote jungles, and about proprietary techniques for grafting. I figured I'd meet indulgent spouses and puzzled neighbors.

And there was all of that. But there was so much more. When you ask people to tell you about the one activity they do not for money, not out of necessity, but to indulge their deepest passions and their wildest curiosities—well, you're in for an intimate conversation.

People told me about their childhoods and their earliest memories of trees. They told me about the parent or grandparent or neighbor who somehow had the foresight to hand an eight-year-old kid a field guide to, say, ginkgoes or eucalypts. Some told me that planting their first tree helped them cope with unbearable loss. Some fell in love on tree-collecting expeditions. Some said their collections helped them to reckon with the past or connect to their heritage. And everyone who plants a tree thinks about the future, and about what lives on after they're gone.

How often do any of us get a chance to pour our hearts out to a stranger? Somehow, talking about trees made it possible. By the end of every interview, I felt like I'd made a friend—but more than a friend, really. These tree collectors—*my* tree collectors, as I came to think of them—seemed more like some distant family I was meeting for the first time.

Chestnut collector Allen Nichols told me that he was related to the chestnut tree through his mother. I felt that I was related to each of these collectors through their trees. I wanted several of them as surrogate grandparents. Quite a few could serve as cool cousins or quirky aunts. People who spoke no English at all—the botanist in Japan, the community organizer in India—came to seem like family, even when all I could do was sit quietly and let an interpreter handle the interview.

You might think that fifty conversations with tree collectors would get repetitive, but I learned that no two tree collectors are alike. They're driven by their own highly individual desires and instincts. Some aim to conserve threatened species, others wish to restore the land, some want to surround themselves with beauty, and some want to make a memorial or create art. Tree planting is a way to renew both the land and the person doing the planting. I started to see that the life of a tree collector is filled with adventure and wonder. It is a life well lived.

It's also, like any other type of collecting, a way of expressing a singular obsession. Collectors tend to find a niche and try to possess every-thing that occupies that niche. This can be a lifelong pursuit. Oak collector Béatrice Chassé told me that this is why she knows so many octogenarians who share her obsession: "Collecting is a passion, and that's what keeps people alive. Maybe it's futile, but I only wish that ev-eryone could have such a futile activity."

Because the reasons for collecting trees are so varied, I decided to group these collectors according to what I saw as their primary motiva-tions. Healers have found a way to heal their own lives, the lives of oth-ers, or even the wounds of the past by planting trees. Artists have forged an art practice through their work with trees. Seekers have taken their passion for trees around the world, or even into space. Community builders have undertaken the remarkable work of knitting people to-gether under an arboreal canopy. It's impossible to reduce anyone's kin-ship with trees to a one-word label, but I hope my classification speaks to the diverse possibilities a life among trees can offer.

As a hobby, tree collecting would seem to pose a major barrier to entry: access to land. Any collection takes up space. But surely no one is greed-ier for land than a tree collector. The story behind many of the world's most extraordinary private tree collections begins like this: "Back in the seventies, we bought thirty acres of land and started planting." But who has the funds for thirty acres of land today? Who was able to afford it then?

These questions led me to think more broadly about collectors and collecting. What about an urban tree-cataloging project? Francisco Ar-jona began cataloging the trees of Mexico City on his Instagram ac-count. Isn't that a collection? Sairus Patel manages the extraordinary

Trees of Stanford project, which not only tracks every tree on Stanford's campus but looks back in time and tracks every tree that ever grew on campus. It's a collection because someone took notice and started counting.

Some collections, of course, take up very little space. You can pick up acorns in your neighborhood and sprout them in pots on your balcony, and I've met collectors who do that. You can graft forty varieties of fruit onto a single tree, like artist Sam Van Aken. Some people collect pine cones, or leaves, or wood specimens. You can store those in a few boxes under your bed.

But confronting the question of who has land and who doesn't—and why—led me to the most meaningful stories in this book.

Joe Hamilton plants trees on land passed down from his formerly enslaved great-grandfather. What started as an 888-acre parcel at the end of the Civil War had shrunk down to 44.4 acres by the time Hamilton and his family inherited it, the balance lost to banks, disreputable deals, and outright theft. Title to the land had never been registered properly. It took a tremendous amount of legal and genealogical research for Hamilton to claim what was left of the land for himself and his heirs. And he planted a collection of just one kind of tree—loblolly pine—to establish long-term, generational wealth for his family through sustainable forestry.

As a college student, Reagan Wytsalucy learned about the Navajo peach orchards that were nearly—but not entirely—destroyed by the United States Army in the 1860s. She was inspired to try to locate the remaining trees and bring the orchards back to life, and now she's building a collection that belongs not to her but to the Navajo people.

A tree collection in Greenland similarly belongs to the public, as does a collection planted to honor girl children in a small village in India. These efforts go beyond the impulse to acquire rare and unusual trees for one's own property, and they speak to the many and varied ways a tree collection can be rooted in community and cultural values.

One other question that arose as I wrote was where, exactly, do these trees come from? Most collectors buy trees the way anyone buys a plant—at a garden center or tree nursery. Some, like Sara Malone, pick up rare trees at swaps or auctions held by a tree society. When a very rare or previously unknown tree is found, such as *Tahina spectabilis,* botanists might organize a program of conservation that could, at some point, include sales of cuttings or seeds to raise funds to help protect the trees in the wild. Encouraging collectors to grow them away from their natural habitat (called ex-situ conservation) can be useful, creating backup copies in case human activity, disasters, or climate change destroy the original.

Nonetheless, there are methods of acquiring trees that might seem reckless, greedy, or downright exploitative. In April 2021, *The Wall Street Journal* ran an article titled "The Newest Status Symbol for High Net-Worth Homeowners: Trophy Trees." (This isn't really the newest status symbol: a nearly identical article ran every five or ten years for most of the twentieth century.) Wealthy people who want the biggest and the best will find a way to get it, and that includes spending hundreds of thousands of dollars to transplant mature trees into their lavish landscapes. Salomé Jashi documented the activities of one such collector in her film *Taming the Garden* and exposed the ways in which poor families in the countryside were left with little choice but to part with beloved trees for a sum of money they simply could not refuse.

But not every tree relocation is a shady act: some trees are moved to save them from demolition during construction, to make room for other trees to grow to their full size, or simply to place them in a more suitable location. Landscape architect Enzo Enea built a museum in Switzerland entirely from trees rescued from construction sites. A friend of mine remembers a man knocking on his mother's front door when he was a kid and offering to buy the palm tree in the front yard. It was a poorly situated tree in a small yard, and his mom was happy to see it moved. She sold him the tree.

There's also the question of collecting seed from remote forests around the world. I spoke to many tree collectors in their seventies and eighties who remember going on fairly unregulated seed-collecting expeditions decades ago. But today, plant hunters are expected to obtain permits and comply with regulations aimed at protecting fragile ecosystems and making sure that collectors from wealthier nations don't walk away with the riches they harvest from poorer countries. These protocols have helped to promote equity, conservation, and the sharing of resources around the world.

Why collect trees? In addition to all the other reasons—curiosity, desire, aesthetics, acquisitiveness, legacy, or community—being around trees feels *fantastic*. The Japanese practice of forest bathing, *shinrin-yoku,* involves fully taking in the forest through all of one's senses for an hour or two at a time. This practice actually changes the levels of stress and pleasure hormones in the body, decreasing cortisol and increasing serotonin.

Tree collectors know this. And if being around one tree feels good, their thinking goes, imagine how a hundred trees would feel.

If this book accomplishes anything, I hope it inspires you to plant a tree. Or two. Or maybe a dozen. Watch out, though—trees can be addictive.

PLANT TAXONOMY

Taxonomists study the genetic relationships between plants. As they make new discoveries, they might reclassify and rename a plant. Some taxonomists like to group several species into one, while others like to subdivide one species into many. Casual observers of this sport often refer to them as "lumpers and splitters."

• A TREE'S FAMILY TREE

Taxonomists would place a sugar maple, *Acer saccharum,* on this branch of the family tree:

Kingdom: Plantae
Division: Magnoliophyta
Class: Magnoliopsida
Order: Sapindales
Family: Aceraceae
Genus: Acer
Species: saccharum

• SPECIES

A group of organisms that share genetic characteristics and are capable of reproducing with other members of the species. As with many scientific terms, this one is not straightforward: some plants can cross-breed between species, for instance. The Latin or botanical name of a plant is its genus and species, such as *Acer saccharum,* the sugar maple. New species are recognized and registered through the International Association for Plant Taxonomy.

• SUBSPECIES

A smaller group within a species that can still reproduce with other members of the species but otherwise share some distinctive physical characteristics, usually because of geographic isolation or different climate conditions. A subspecies name is written with the abbreviation *subsp.,* like this: *Acer saccharum* subsp. *nigrum.*

• VARIETY

A naturally occurring variation within a species or subspecies, usually due to a noteworthy physical characteristic like flower color. To the great frustration of tree collectors, varieties are often reclassified as subspecies and vice versa. For instance, the black sugar maple, named for its dark-gray bark, is sometimes described as *Acer saccharum* var. *nigrum.* Varieties usually grow true from seed, meaning that the plant's offspring will carry on those characteristics.

• CULTIVAR

A cultivar, short for "cultivated variety," is a plant that has been bred by humans for specific desirable qualities. Cultivars will not grow true from seed, meaning that plants grown from their seeds will not look like the parents. Cultivars generally have to be cloned, through grafts or cuttings, in order to reproduce. Cultivar names appear in single quotes after the species name, like this: *Acer saccharum* 'Green Mountain'. New cultivars are registered and recognized through that plant's International Cultivation Registration Authority. For example, maples are registered through the international Maple Society. Cultivars can also be patented, like any other invention.

xxvi • Tree Terms

• SELECTION

A single specimen of a plant that has some desirable characteristic, such as faster growth or an earlier blooming season. (Think of a rancher who selects a particularly good bull for breeding purposes.) A selection can be cloned by a grower or nursery to offer a slightly superior version of a species, subspecies, or variety.

• HYBRID

The offspring of two different species, subspecies, or varieties, indicated with an × in the name, such as *Acer griseum* × *saccharum,* also known by its cultivar name, 'Sugarflake'.

• SPORT

A genetic mutation that causes a new or different physical characteristic to appear on part of a plant, such as a tree with one branch of variegated leaves.

GRAFTING

The act of cloning a plant by attaching a twig or shoot from one plant to another.

ROOTING

The act of cloning a plant by cutting off a stem or twig and placing it in water or soil so that it can grow new roots.

MONOECIOUS

A plant that produces both male and female flowers, so that it can reproduce without another plant nearby. For instance, birch and spruce trees are monoecious.

DIOECIOUS

A plant that produces either male or female flowers on separate specimens, so that two plants must be in close proximity to reproduce. Maple and aspen trees are dioecious.

WHAT'S THE DIFFERENCE BETWEEN A TREE AND A SHRUB?

Trees and shrubs both have woody stems, but trees grow on a single stalk or trunk, while shrubs produce multiple trunks. However, some shrubs, like rhododendrons or camellias, might take the form of small trees. And some tropical plants, like bananas, are commonly called trees even though their tall, single stalk is not woody.

THE TREE COLLECTORS

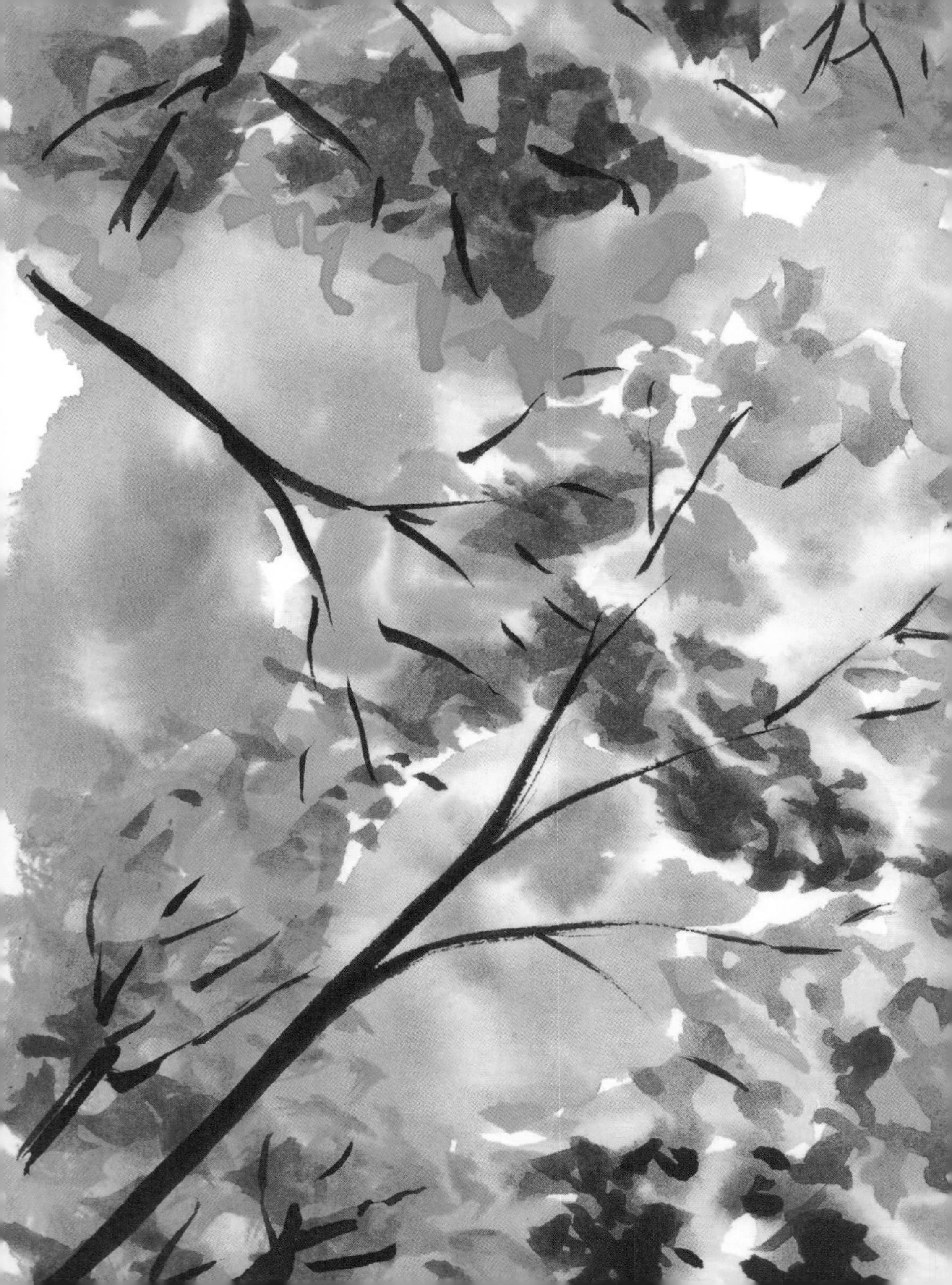

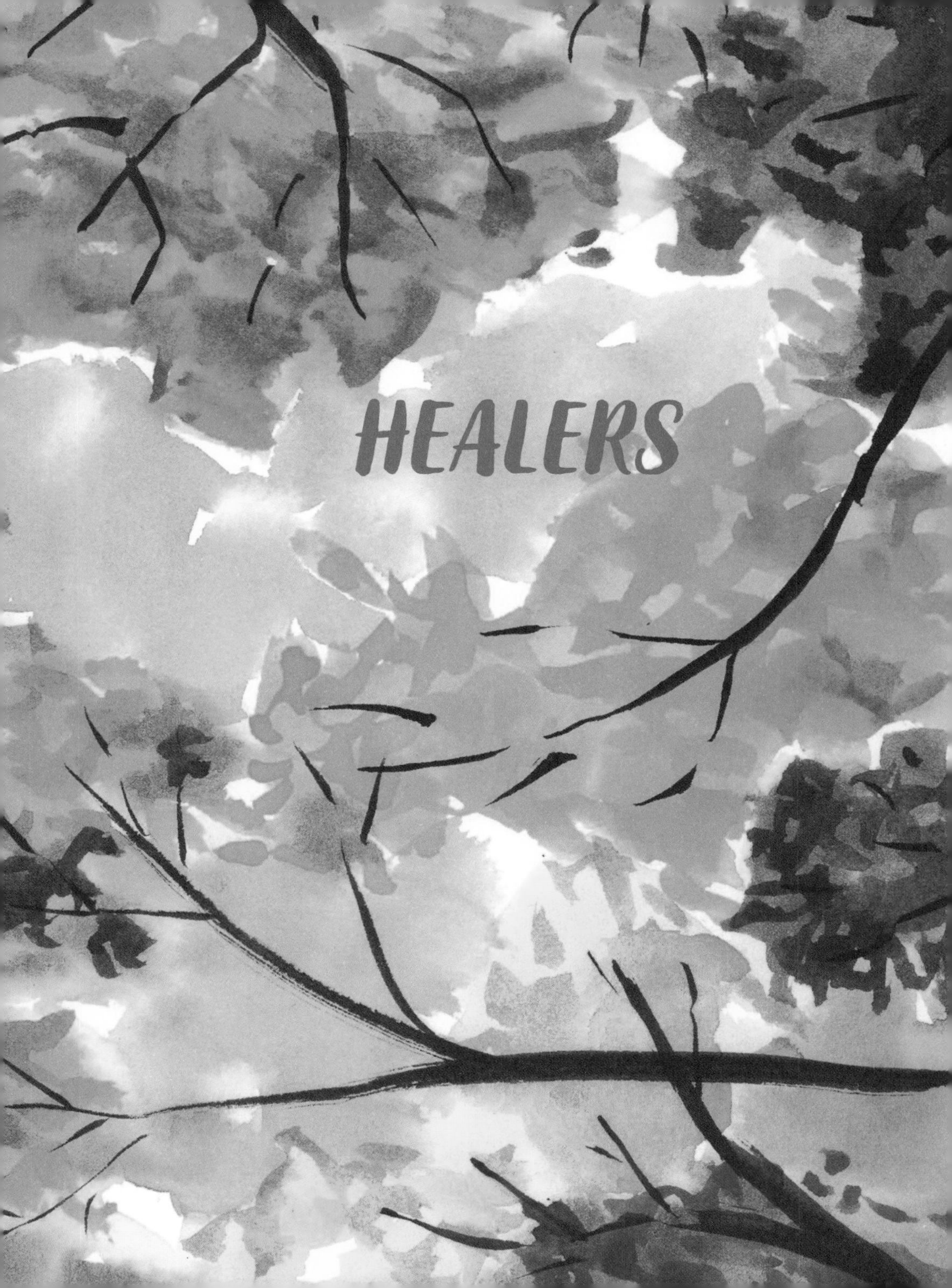

HEALERS

"GROWING FRUIT TREES IS A VERY SIMPLE WAY TO STAY IN LOVE WITH OUR WORLD."

THE PLAYWRIGHT

VIVIAN KEH

San Jose, California

WHEN VIVIAN KEH WAS A STUDENT AT THE YALE SCHOOL OF Drama, she wrote a play called *Persimmons in Winter*. "It was about two Korean sisters who survived World War II and the Korean War," she said. "It was based on my mother's experience. She went through some very hard times, times of starvation and war. The metaphor is that the sisters are the persimmons. It has always seemed miraculous to me, the idea of a tree producing fruit in winter."

She planted her first persimmon tree in 2012, after she and her husband moved to a suburban home on a quarter-acre lot in San Jose. Persimmons are categorized into astringent and non-astringent types; she grew both the non-astringent 'Fuyu' type, which produces flat, squat fruits that can be eaten when they're still firm, and the astringent 'Saijo' variety that can't be eaten until fully ripened. "Those are the ones my elders are familiar with," she said. "You bring them home from the market and then wait until they get really soft before you eat them. They remember these from when they were young. This was their sweet! They would also eat them dried with a cast over the skin from all the sugar coming out. That's a delicacy. And there's something about feeding oneself something sweet when you've been through starvation. It means a lot to them."

Vivian recalls that in Korean culture, persimmons are a Buddhist symbol of transformation, shared in celebrations and placed on altars and grave sites to honor the dead. But to her, persimmons signify her connection to nature and to her family.

One fruit tree led to another, and now she has fifty trees, including citrus, quince, apricot, and medlar, an apple relative that also can't be eaten until it's so soft that it almost appears rotten. But the persimmons are the centerpiece of her collection, and a sort of spiritual force in her home orchard.

"The 'Saijo' persimmon I planted—there's something special happening around that tree," she said. "There's energy around it. I feel like there's some connection to my ancestors, to the ones I never knew, even to the ones who've been forgotten. All I know is I feel really good when I'm hanging out around that tree. I talk to it, and I thank it."

The experience of living with these trees has helped her to reckon with the past, and with the whispered recollections of her elders that she struggles to understand. "I think maybe a couple generations back, some hard choices had to be made. Choices like, do I save my son or my daughter? What I think happened, maybe in my great-grandfather's generation, is that two girls were left to die. And even though I don't know their names, I think about them. Because they mattered. They're the ones I talk to when I talk to that tree. It might sound crazy, but you know, that branch, the original branch that all these grafts came from, it came from Asia. And here we are together in my backyard."

Every winter, as the persimmons ripen, she packs up boxes of them and mails them to her relatives. "Being a child of immigrants, relationships can get complicated. One way I feel like I can offer a gesture of love is by giving my elders these fruits that they adore. It keeps me connected to them. Now it isn't enough to send one box to my mother. She wants more! I'm sending her three boxes a year."

It also connects her to a community of passionate fruit tree growers. "You see these T-shirts that say 'Introverted but willing to discuss books'? That's me. Introverted but willing to discuss fruit trees." Every year, she meets up with other fruit tree enthusiasts for an exchange of scion wood, the thin branches used for grafting. Thanks to those exchanges, she's grafted as many as fifteen persimmon varieties onto one tree.

"I love starting with small trees and just pruning them and working with them," she said. "They grow with you, and you help them to grow into the beautiful shapes they have the potential to become. I want to grow old with these trees, so I keep the canopies low. I hope to keep harvesting this fruit into my eighties. Fruit trees are just amazing—they give us so much for so little."

Lately that good feeling has mattered more than ever. "There's so much that really disgusts me about our society right now," she said. "There's all this Asian hate crime that we're seeing. It's really depressing. But when I harvest these persimmons, and I put them on the cutting board and start cutting . . . it's such a pleasure. And I start singing! They really do make me sing. Growing fruit trees is a very simple way to stay in love with our world."

"THE TREES HAVE TAKEN ON THE PERSONALITIES OF THE PEOPLE THEY'RE ASSOCIATED WITH."

THE MEMORIALIST

LINDA MILES
Netherton, England

WHEN LINDA MILES AND HER HUSBAND BOUGHT A PLOT OF LAND in Herefordshire, back in the midseventies, it was all farmland. "There wasn't a tree in sight," she said. "But a friend who was a tree enthusiast was moving away and gave us some rare conifer seedlings to look after. So we began making space on the land to plant some trees."

It was some time before they realized what a gift they'd been given with those first seedlings. "We were on a maple society tour one time, and our guide pointed to a Bhutan pine and said, 'This is really rare,' and I thought, *Oh, we have fifteen of those.*"

Gradually, they began to fill thirteen acres of land with rare and unusual conifers and maples. They were raising four children and working full-time—she was a geography and geology teacher, and her husband, Tim, was a concrete engineer—but when they could, they traveled through Europe and Japan to find trees.

"We've planted over fifty species of maple, all grown from seed," Miles said. "The Chinese paperbark especially is just exquisite, with that cinnamon-colored bark and bonfire-red leaves. I'm still after *Acer erianthum,* which comes from China and puts out reddish-purple fruit in the summer, but it's difficult to raise from seed and I've never managed it."

Collectors tend to make the rookie mistake of planting trees too close together, but Miles went in the opposite direction. "What I really wanted was to appreciate each individual. You get used to seeing trees bunched together, and I know they can give marvelous color that way, but I just love the natural look of trees. I set them each apart from the others, so they can grow into their own shape. They aren't pruned at all, unless a branch dies back."

As she walks through her trees—all loosely grouped into islands with wide grass paths between them—it isn't so much the rarity of any particular tree she thinks about, but the person associated with it.

"Just as we were coming to view this place for the first time, our dear friend Michael was killed in a car accident. So we didn't visit the property for about three weeks. But then, when we did move in, one of the first trees we planted was a very beautiful, very regal-looking cedar of Lebanon in Michael's honor. His children have come back to see it. It's

just huge now, with these lovely horizontal branches, and it reminds us of him."

That first memorial tree set the tone for what is now a kind of community of trees, all planted to mark an occasion, celebrate a milestone, or remember a loved one. "With every tree, there's an association of thought," she said. "We have four children, and their spouses, and ten grandchildren, and they each have a tree. I'm always amused because I look at them, and the trees have taken on the personalities of the people they're associated with. My daughters-in-law are strong and vibrant and beautiful, and so are their trees. I just love to go out and look at them. And the trees smile back at me."

Any occasion can provide an excuse for a tree planting: over the years, friends have been invited to plant a tree when they visit. Miles has brought potted trees to church to decorate a wedding, which the couple planted after the ceremony. Her son-in-law built a treehouse for the grandchildren in the arms of two enormous conifers. Now an extended fellowship of friends, family, and neighbors are all connected to one another through the trees.

"My daughter and her friends from university came down at the millennium, to celebrate the new year, and they dressed up for the evening and decided they'd like to plant a tree. So we took our torches up to the location that I thought was appropriate, and they planted their millennium tree. Now it's twenty-two years on, and it's lovely. A couple of them still come to visit, and they've planted trees for their children as well."

For the last twenty-plus years, she's carried on planting trees without her husband, who died in 2002. "For the first couple of years after he died, I didn't feel like doing much. And then I thought, *Right, let's get on with it.* I keep this saying on the wall: 'One who plants a tree looks at tomorrow, not today.'"

"HE TOOK ME ASIDE AND QUIETLY SAID…'I DIDN'T KNOW THE WORLD COULD BE SUCH A BEAUTIFUL PLACE.'"

THE ARBOREAL THERAPIST

JANUSZ RADECKI

Pruszcz, Poland

Janusz Radecki's tree collection isn't located at his house, nor can it be found at a public arboretum. His collection is planted on the grounds of a nursing home for people with chronic mental illness in Gołuszyce, a quiet village in northern Poland.

He'd been trained as a fine artist, and for five years, he ran a sign-painting business, but then the work dried up. "There were no more orders, neither for signboards nor for paintings, and I simply had to look for another job," he said. "My wife was working at the nursing home, and she recommended me to management."

He worked in administration, and then taught art. "Unfortunately, there weren't many people at the nursing home who were interested in these types of things," he said, "but there was a group of residents who really enjoyed walking in our grounds."

He was already an enthusiastic gardener. With the help of a couple of friends, he developed a horticultural therapy program for the residents, the first of its kind to be introduced to a nursing home in Poland. "Horticultural therapy can be divided into active and passive types. I do both, although I emphasize the active one. Our type of therapy is that we physically take care of the plants—we rake leaves, we plant trees, we water, we weed." The passive element matters too: "The fact that the patients can be in a beautiful environment of plants is also of no small importance. It has been scientifically proven that being surrounded by greenery improves psychological comfort."

The program wasn't an immediate success. Management didn't sup-port the idea because, back in the 1990s when Radecki began, there wasn't much research on horticultural therapy. And employees began to steal and sell newly planted trees. "It was another five years before we took care of the problems and had a good program," he said.

He met plant breeders and owners of tree nurseries, who were happy to provide interesting and unusual varieties of trees for his program, many of which are not found anywhere else in Poland. A magnolia breeder in New Zealand, Vance Hooper, introduced 'Genie', which had an excep-tionally long blooming season. "Most magnolias bloom for only a few days," he said, "but this one has a dark cherry-beet color, blooms in spring for nine weeks, and then blooms again in August."

American maple breeder J. Frank Schmidt offered a maple, *Acer ru-brum* 'Redpointe', with fast growth and an exceptionally fiery red color in the fall. Other growers have supplied dogwoods, flowering cherries, rhododendrons, and witch hazels. Radecki tries to run a year-round program for the residents—or as close to year-round as Poland's formi-dable winters will allow—so having a vivid, spectacular show of flowers and foliage from early spring to late fall is a priority.

He's now planted more than eight acres with trees and shrubs for the residents to care for. There's a great deal for them to do—and that's the real advantage to his program. "Another facility in Poland has every-thing you can dream of, but the residents can only watch it," he said. "They don't work with plants. They don't take care of them as we do. I have a friend near New York City who works as an occupational therapy instructor. He envies the fact that we can take care of our plants our-selves and that no one imposes anything on us. I think that our nursing home differs from others in terms of liberty. We decide ourselves what we are going to grow, where, and in what size."

Even being in the presence of trees can be transformational for the

residents. Radecki remembers a young man who was referred to the institution by court order. He didn't want to go on any field trips or participate in any activities. He finally agreed to join an excursion to Rogów Arboretum, about a three-hour drive to the south.

"I saw this young man walking around this arboretum with an open mouth. I could see how much of an impression this place made on him. Then he took me aside and quietly said, 'Thank you very much. I didn't know the world could be such a beautiful place.' That was one of my greatest therapeutic successes."

"WHEN YOU LOOK AT THE VOLUME
OF LAND THAT WAS LOST, AND
EVERYTHING WE WENT THROUGH—IT
WAS NOT IN VAIN. BECAUSE NOW
IT'S HERE FOR THE KIDS."

THE RIGHTFUL HEIR

JOE HAMILTON

Green Pond, South Carolina

JOE HAMILTON REMEMBERS WATCHING LOGGING TRUCKS DRIVE past while he waited for the school bus. "These massive trucks were roaring by, and I wondered how they got the trees on those trucks. It did not click with me until a few years ago that those were loblolly pines. And here I am now, with all these pine trees right outside my window."

Hamilton grew up on land that his father had inherited. "My father was a row crop farmer, and he also did labor on other people's farms, kind of like a sharecropper," he said. What Hamilton didn't know at the time was that the land his family lived on had been passed down through generations without their ever having clear title to the property. This arrangement, referred to as heirs' property, has been a particular problem for Black families who have inherited land from formerly enslaved ancestors.

The situation was never explained to him when he was young. "I was only five or six years old when a man came riding up on a horse to talk to my father. He said, 'Well, Steve, you know it'd be a good idea if we could maybe join this place back together.' I didn't know what he was talking about, join it back together! My dad just said, 'Well, no, sir, boss, I don't know if that's a good thing. I don't know what my children are going to do with it but I'm going to hold on to it.'"

The mystery continued into adulthood, when Hamilton started to receive the property tax notices on the place. "The envelopes are always addressed to the care of the property owner. You see? The county

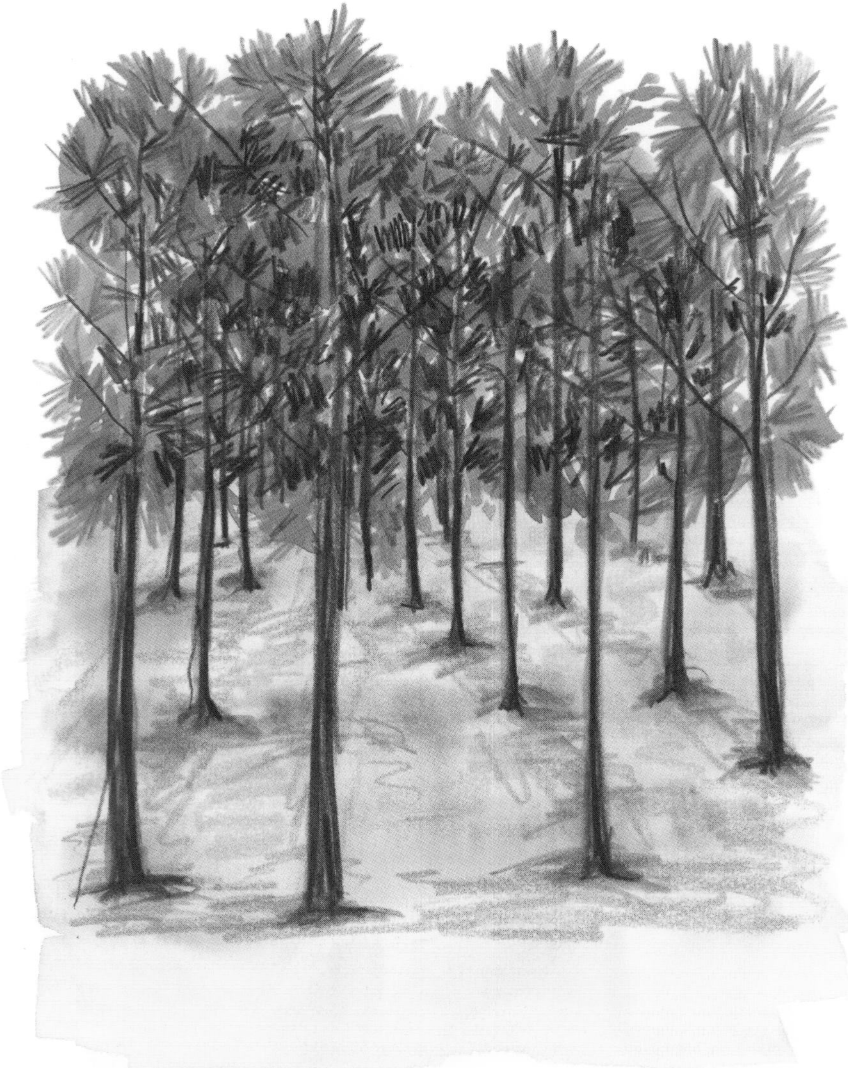

doesn't know exactly who owns it, either. Finally my wife said that we had to get this sorted out."

A local nonprofit, Center for Heirs' Property Preservation, helps South Carolina families sort through the complex paperwork and extensive research required to correctly secure title to their land. They also support landowners who want to put their land to work by engaging in sustainable forestry. In many cases, this is the best use of rural land: two-thirds of the state is forested, and timber is one of its most important industries. If it's done well, sustainable forestry can preserve biodiversity, mitigate carbon emissions, and help reduce the demand for logging in wild forests and jungles around the world.

But before Hamilton could think about any of that, he had to untangle the complex history of his father's land. "I found a second love," he said. "My first love was my wife. My second love is history."

What he eventually learned was that the 44 acres that he, along with other relatives, had inherited was once part of a much larger 888-acre parcel owned by his great-grandfather, Stephen Cunningham. "When I did the research to clear the title, the last known tax document was for 345 acres. So the slave owner turned over 888 acres to my great-grandfather, but he couldn't hold on to it."

Hamilton looks back on those days and tries to imagine how his ancestor must've fared. "After slavery, in Reconstruction, it was a horrible time. Stephen Cunningham had always been told where to go and what to do. His day was governed by someone else, like a prisoner. And now, the slave owner conveys this property to a guy who was himself considered property at one time. But he didn't know how to manage it. He didn't know what to do. He lost land. He tried to barter acreage for a cow and a calf. He sold some land, and he mortgaged some land against his crops. But if you didn't produce the crop, you defaulted. So he lost some land that way."

If Stephen Cunningham managed to hold on to 345 acres, what happened after that? Hamilton isn't sure. "We have 44.4 acres. My endgame was to make sure we had that land locked down. I know a significant bit of land was taken. I can't prove it, and I was exhausted from the years of research. This is what I could do. I would love to have all of it back. But I'm doing okay."

In his research, he learned that one of slavery's fiercest defenders, Senator Robert Barnwell Rhett, owned land that now borders his own. "If he was still living," Hamilton said, "we would be neighbors."

Now that ownership has been established and the land properly divided among the heirs, Hamilton can turn his attention to a long-term timber management plan on his portion using loblolly, a fast-growing pine that can yield good lumber, telephone poles, and building joists.

Most tree collectors wonder who would possibly want to inherit their collection. But this collection—of just one tree, planted over and over—is intended specifically to create generational wealth and continuity. Hamilton's adult children are already well established in their professions, and he's looking ahead to his grandchildren. "I just want them to have some connection to the land. Whatever they pursue as a career, I want them to know that this is for them. When you look at the volume of land that was lost, and everything we went through—it was not in vain. Because now it's here for the kids."

"I'VE NOT BEEN ABLE TO GIVE A NAME TO A BABY. I WOULD LIKE TO NAME A TREE."

THE MOTHER OF TREES

MARIE NOELLE BOUVET

Ipswich, England

WHEN MARIE NOELLE BOUVET'S FAMILY MOVED FROM FRANCE to New Zealand, she was only seven and didn't speak English. "My teacher had me cut pictures from magazines and write the English words. I made a beautiful scrapbook, and it became my first dictionary."

That started a habit of collecting images. "I collected stickers and postcards, and then badges and pins. I had boxes and boxes of them. When I grew up, I couldn't be bothered with them anymore and I started collecting clothes. I wanted every fabric and every color."

But all that accumulating started to wear on her. "What can you do with three hundred shoes? Nothing. I wanted something that would satisfy my curiosity, and the pleasure of the hunt. But all these things, they only remind you of where you've been before and what you did. I wanted something that would connect me to the future, not the past."

The answer came in the form of an abandoned Japanese maple (*Acer palmatum*) seedling in a pot, left behind by a previous tenant at a flat she and her husband were renting. "We carried it from house to house," she said, "and when we finally had a place with a third of an acre, we planted it."

That first tree was the start of something transformational. For some time, she had been carrying a sense of loss because she and her husband were unable to have children. But she had an inkling that planting trees might help her through her grief. She grew fifty Japanese maples in that first garden and took real pleasure in nurturing them along. But there

was one problem: she didn't know their names and couldn't identify them. "We were finally able to move to a nineteen-acre property, and I told myself that I would not add a tree to the collection if I didn't know its name. I had to memorize them. I would say their names in the car, or going to sleep, or walking my dog, and try to picture each one with its name. It became my mantra."

Today her collection has grown to four thousand trees, representing about 650 species of maple. "If it has the name *Acer* in front of it, I have to have it."

Her work as a veterinarian keeps her at home in Ipswich, near the Norfolk and Suffolk border in England, so she's rarely able to travel to find unusual maples. But her veterinary practice gives her another advantage. "I have to import medicines, so I'm quite familiar with customs paperwork," she said. "I've started to import maples as well, and to sell the extras."

Now she hopes to introduce a new maple variety of her own. "With maples, the seeds are never true to their parent trees. You don't know what you'll get. I keep sowing the seeds and looking for one that will be spectacular. I think about what I would name it. I've not been able to give a name to a baby. I would like to name a tree."

The process of introducing a new tree cultivar is a formal one that requires registering with an international organization like the Maple Society. Some people take additional steps to patent or trademark the plant. Bouvet hopes to do this with one of her own seedlings someday, but until she does, she's informally bestowed names on some of them.

"The trees have become a kind of family to me," she said. "They filled a void in my life left behind because I never had children. Somebody once said to me that I'm 'sterile-minded' to think that a woman couldn't be a mother, or would never be a mother, because she didn't give birth. And that resonated really deep in me. The last thing I want is to be sterile. This is why I've planted all these seedlings. I see them as my family."

"I WISH SOMEBODY COULD DO THAT ON THE INSIDE FOR ME."

THE LEAF MENDER

ALPANA VIJ

Singapore

A TREE COLLECTION REQUIRES A CERTAIN AMOUNT OF LAND. BUT a leaf collection? That only takes up a little space on a bookshelf or a wall. Botanists and naturalists have been pressing leaf specimens between the pages of books for centuries. Amateur collectors can create a private herbarium, where leaves are preserved, mounted, and labeled. A leaf collection can even become a kind of artistic practice.

Visual artist Alpana Vij collects leaves in her neighborhood in Singapore. "In 2016 I started taking a walk and picking up fallen leaves. It was almost like a walking meditation. You're immersing yourself in a task, and ideas are kind of coming through because your ego is at rest. You're just catching life as it flows."

Then she started to notice the minute damage on the leaves she picked up: the holes left behind by insects, the cracks and tears, the spots and blemishes. "I was just called to repair those leaves, to stitch them back together," she said. She was a painter, and she'd never before worked with thread. She found a supplier of vintage Japanese silk thread wrapped in gold leaf, and she started to make tiny, intricate patches on the leaves.

"I was quite attracted to this Buddhist concept of *Śūnyatā,* which loosely translates to emptiness, but what it really teaches us is that nothing has a permanent or independent existence. Things only exist in interconnection with everything else. To give an example of a seed, we know it as something that grows into a plant, but actually it's the soil, the sunlight, the rain, everything comes together to birth the tree. There's also this constant change, from the seed growing into a sapling, and then into a tree, that shows that none of the states are permanent. Things are always in flux."

She was also drawn to the Japanese principle of *wabi-sabi*, an observation and appreciation of imperfections, and to the art of *kintsugi*, in which broken pottery is repaired with gold, so that the process of mending is celebrated as part of the history of the object.

Her stitchwork on the leaves is dazzling in its precision and its brilliance. It's hard to imagine a human hand making such delicate repairs, but every stitch conveys a startling tenderness. Some of the leaves are too brittle, so she makes a tiny patch with another leaf and marries the two together.

She mounts the leaves on paper that she makes herself, or against concrete blocks from a local concrete fabricator, then seals them in a climate-proof frame. Each one is titled by the date and road where the leaf was found. She exhibits them in galleries in Singapore and in India, where she was born.

Although she collects only leaves that have already fallen from the trees, she did start to notice damage on leaves still attached to branches. "I started bringing my needle and thread with me on my walks, and I would repair the leaves on the trees," she said. Those tiny, fragile works of art would remain where she found them, suspended from branches. "I never said where they were, but some of my friends would see them on Instagram and go looking for them."

Vij views her art practice as a study in fragility, strength, and interconnection. "It's quite a healing act, to bring a leaf home, and wash it and soak it and dry it and mend it," she said. "A friend of mine said, 'I wish somebody could do that on the inside for me.'"

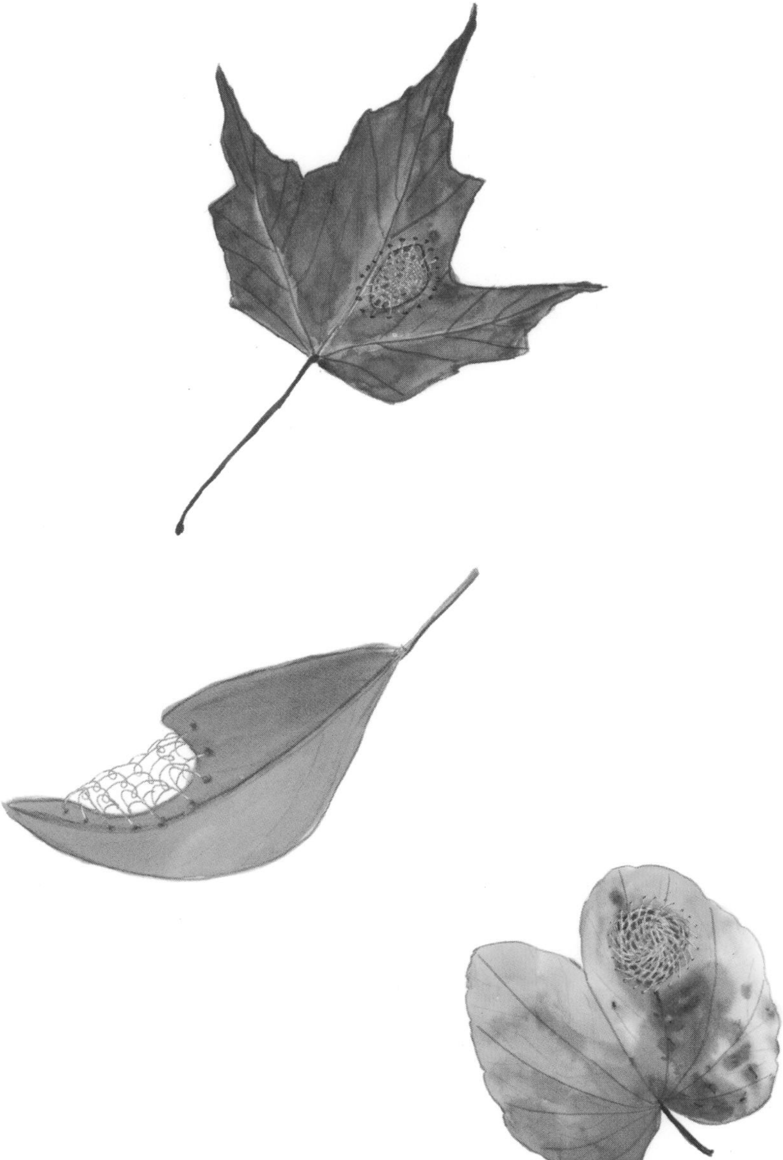

HOW TO MAKE A LEAF COLLECTION

1. CHOOSE A THEME

Every tree on your street

All the oaks

Just one tree all year long

Color-coordinated

2. COLLECT THE LEAVES

Get permission and know local rules

Wipe off dirt, bugs, etc.

*Take notes—
what/where/when*

3. PRESS AND DRY LEAVES FOR FOUR TO SIX WEEKS

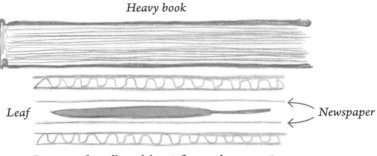

Heavy book

Leaf

Newspaper

Corrugated cardboard for airflow and evaporation

4. MOUNT AND LABEL

PVA GLUE

Sketchbook—smooth, heavy paper

Dab glue on back of leaf with brush or sponge

Wax paper

Weigh page down until glue dries

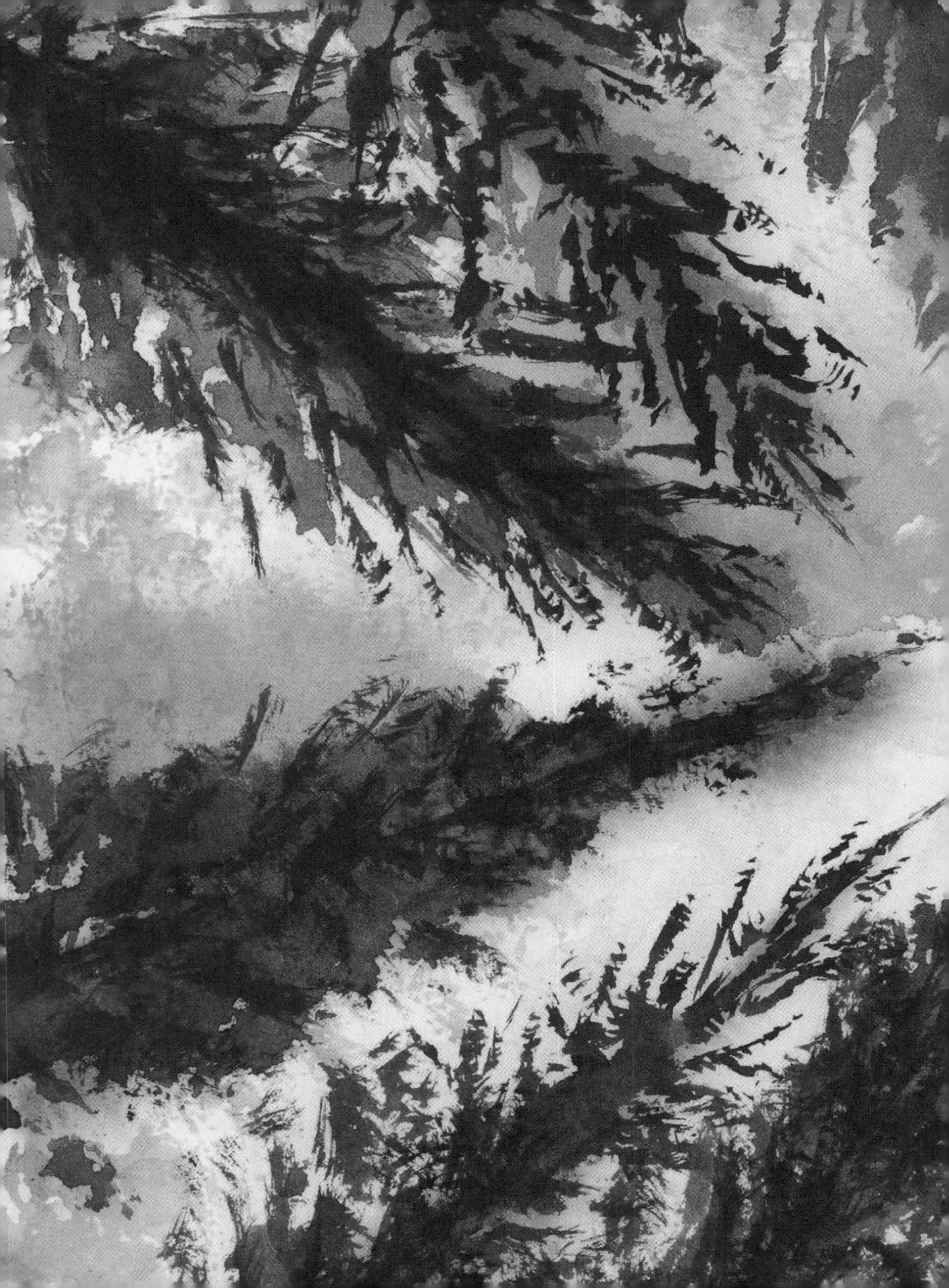

ECOLOGISTS

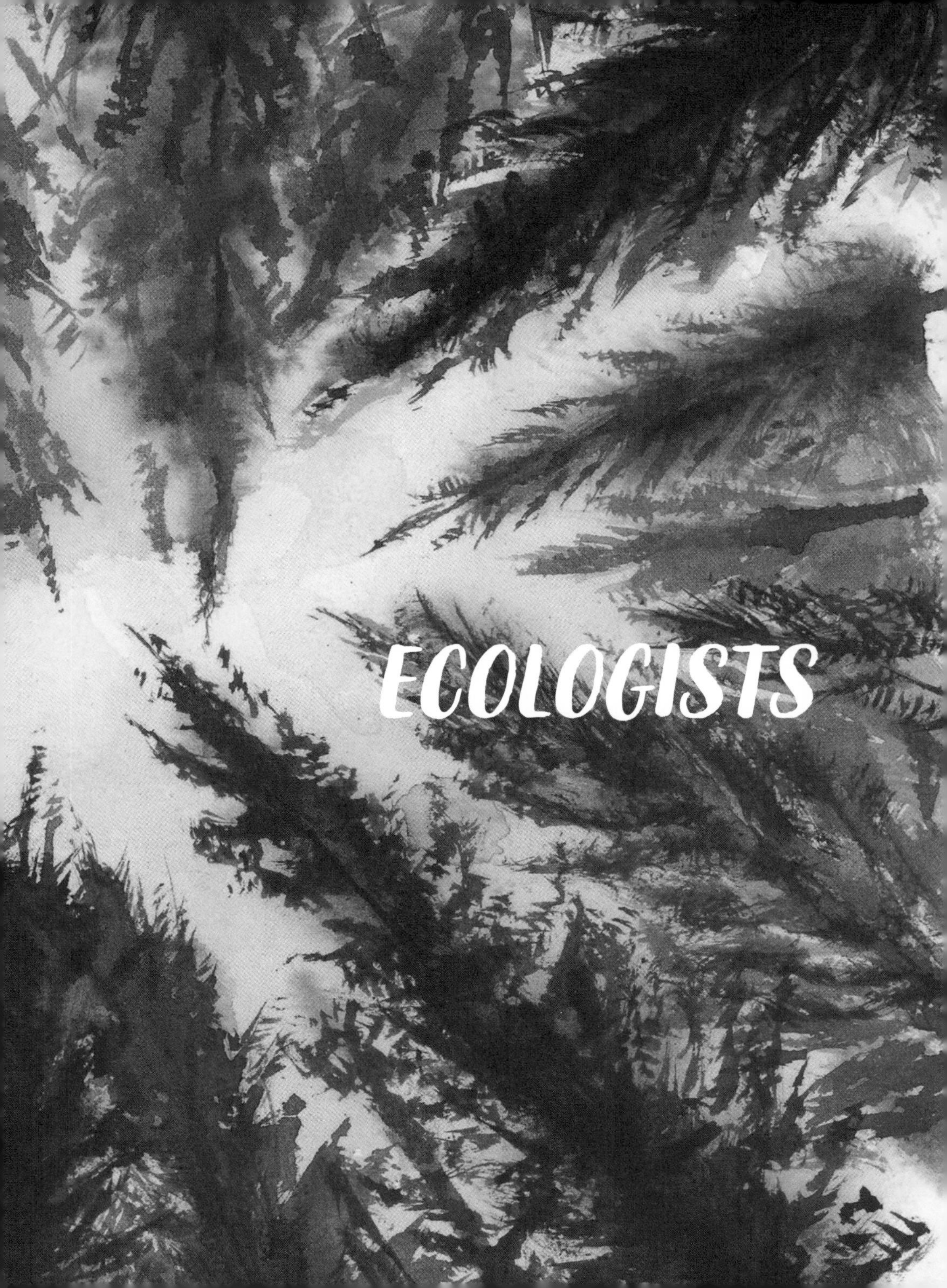

"YOU CAN'T IMAGINE WHAT IT'S LIKE,
TO NEVER SEE A TREE."

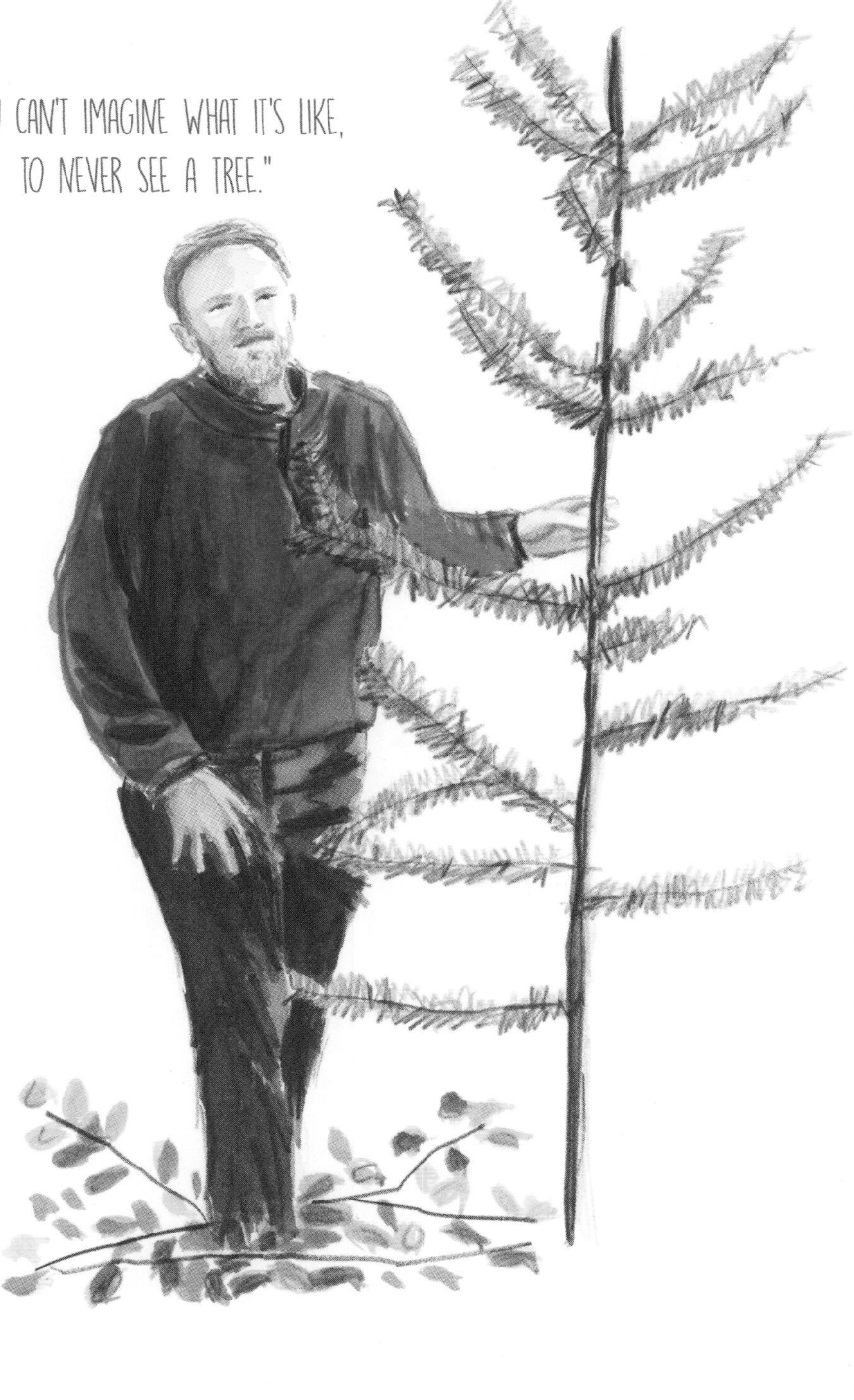

THE ARCTIC ARBORIST

KENNETH HØEGH

Narsarsuaq, Greenland

KENNETH HØEGH GREW UP IN A TREELESS PLACE. "WHEN I WAS thirteen or fourteen, I found a book in the library about agriculture in Greenland. There was just a small section about tree planting. I asked my father if we could plant a tree. He was always very supportive of anything I wanted to do, so he asked the agricultural station for help, and they were able to find four Siberian larches for us."

Greenland's location along the Arctic Circle gives it a polar climate that has not, until recently, supported tree life. The scant forests that could survive in southern Greenland had been cut down for timber, and the land where they'd stood was given over to grazing. Høegh, the son of a mixed European and Greenlandic-Inuit family, grew up in the southern town of Narsaq, where he remembers sheep walking around town and grazing freely. "It wasn't easy to even find a tree to plant in our garden," he said. "We just didn't have them. You can't imagine what it's like, to never see a tree."

He studied agriculture at the University of Copenhagen and wrote a master's thesis on the subject of tree planting in Greenland. That led to regular summer tree-collecting expeditions with botanical research groups. The goal was to find trees that were growing well near the arctic tree line, which is the northernmost boundary where trees will grow. Alaska, British Columbia, Manitoba, and Siberia all offered interesting possibilities.

Høegh has had a career as an agricultural adviser and a diplomat, representing Greenland to the United States and Canada, but his tree work continues at home in the summer. He helped to establish the Greenlandic Arboretum in Narsarsuaq, which now spans 370 acres and hosts more than 125 species, subspecies, and cultivars of trees and shrubs. A warming climate offers a longer growing season and the possibility of establishing birch, larch, and conifer forests throughout southern Greenland.

Today birds, including cedar waxwing and fieldfare, are starting to move into the forests, and an entire ecosystem is springing up around the trees. Høegh and his colleagues don't know yet whether the arboretum will be used for botanical research, timber production, carbon sequestration, or as a wildlife sanctuary or a public park. "That's for the next generation to decide," he said. "We are taking the first step, which is to find out what trees will grow in Greenland."

He has his favorites among the species that have taken hold. The alpine larch, *Larix lyallii,* is a particularly stunning conifer that glows a brilliant, deep gold in fall. "People can't believe that color when they see it," he said. "It's the most hardy conifer for our cold climate, and it can live a thousand years." Bristlecone pine, *Pinus aristata,* is also making a home for itself in Greenland. It thrives at high altitudes and in the coldest climates and puts out small deep-purple cones when it's ready to set seed.

The arboretum has already accomplished one thing: it has given Greenlanders a place to come and be among trees. "Trees make life livable," he said. "We live in this treeless place. People feel so happy going into a forest, to have all the smells of a forest, and the sound of the wind in the trees. Now they want to plant a tree in their own garden, and we can tell them what will grow. It's a beautiful thing."

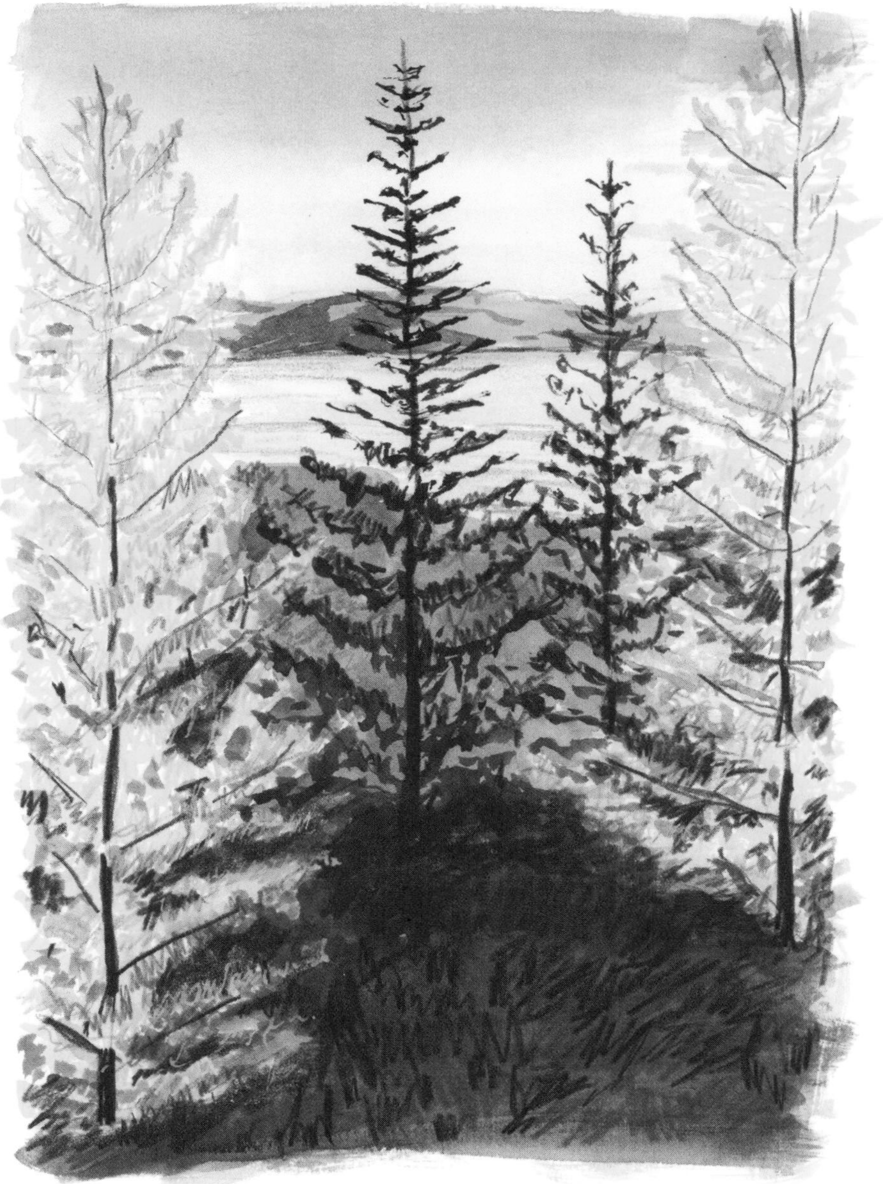

"A FOREST AT YOUR OWN HOUSE IS PROBABLY THE GREATEST LUXURY ANYONE CAN HAVE."

THE TINY-FOREST ENGINEER
SHUBHENDU SHARMA
Uttarakhand, India

SHUBHENDU SHARMA STUDIED HIS ENTIRE LIFE TO BE AN ENGI-neer. Then, just six months into the job he'd always wanted, at a Toyota plant in Bangalore, he met a botanist who would change the course of his life.

But even before they met, he'd started to feel that something was wrong with the career path he'd chosen. "I went into supply development. Our role was to understand the entire process of how a tire is made, or a small part like a nut or a bolt. We would go to the suppliers, and then to their suppliers, until we got to the source of the raw materials. And I started seeing that almost everything starts from a natural resource, and it all ends up in a junkyard. There is nowhere else for it to go but the junkyard."

An engineer thinks in terms of systems and can chart the logical outcome of a sequence of steps. What he realized, when he considered every step in the process of making a car, was deeply unsettling to him. "Is it for the good of humanity that we're making ten million cars every year? Or is it because we want to keep our jobs, keep our machinery running? Because one day, maybe it'll be the hundredth generation, there won't be anything left to convert from a natural resource to end up in a junkyard."

That's when he met the Japanese botanist Akira Miyawaki, who had been hired by Toyota to plant one of his tiny forests at the factory's campus in Bangalore. "My boss said that somebody from Japan had come to give a lecture on the environment," Sharma recalled. "He was not in a good mood that day, so he said sarcastically, 'So who is going to go and attend it?' My job was very hectic, and I wanted a break, so I said I would go."

Miyawaki had been invited to explain his method of planting a dense, rapidly growing, self-sustaining forest on any small plot of unused land. An area the size of a few parking spaces would work, although a typical tiny forest is about the size of a tennis court. His ideas stood in sharp contrast to the industrial production cycle Sharma was starting to question.

"If you plant a forest today and help it to grow for the first two or three years, and then go away and come back after twenty years, you'll see that place flourishing with life," Sharma said. "That cannot happen with any other industry. Why? Because when you're working with a forest, you're working with nature. In any other industry, we're continuously working against nature. You want to save something from corrosion? You're always sanding it, putting primer on it, maintaining it. Why? Because corrosion is a natural phenomenon. You have to fight it."

Sharma volunteered to help Miyawaki plant the forest at the Toyota campus. The method had been rigorously tested and systematized. It involved densely planting local native species in a plot of land that had been intensely cultivated and inoculated with soil microbes. Nothing more was required, beyond heavy mulch and a little supplemental water for the first few years. The roots would form a massive web, the trees would grow rapidly, and within a few years, the forest would be practically impenetrable, making it the ideal host for insects, birds, and other wildlife.

Once Sharma saw the method in action, he tried it at home, installing a tiny forest in his backyard. "This was so much more joyous than doing anything else. I could not get this idea out of my mind, that this was something I should be doing for the rest of my life."

In 2010—just three years after that fateful meeting with Miyawaki—he quit his job. Today he installs tiny forests around the world and teaches people how to do so in their own backyards. As a result of his work, at least 4.5 million trees have been planted in forty-four cities across North and Central America, Europe, the Middle East, and India.

Sharma emphasizes that planting a Miyawaki forest is no substitute for conserving ancient, wild forests. What he does is not reforestation but afforestation, a process of planting a forest on land that is currently treeless. These pockets of forest, whether they're installed on corporate campuses, in city parks, in unused areas around freeways or airports, or behind someone's house, can still behave like natural forests by supporting wildlife, sequestering carbon, and controlling erosion.

They're also beautiful, especially in a backyard. Even a tiny forest evolves as the shrubs and smaller trees give way and the larger canopy trees mature. Leaves fall and build a new layer of mulch. Small fruits and nuts attract wildlife. Birds build nests, and caterpillars give way to butterflies. It's an ever-changing natural panorama. "I think having a forest at your own house is probably the greatest luxury anyone can have," he said.

HOW TO PLANT A TINY FOREST

The Miyawaki method, systematized by Shubhendu Sharma:

1. Survey local forests for native species. Look for shrub, subtree, tree, and canopy tree species.

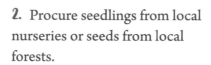

2. Procure seedlings from local nurseries or seeds from local forests.

3. Dig soil at least three feet deep and incorporate locally sourced materials to add nutrition, perforation, and water retention (examples: manure, rice husk, and coconut fiber).

4. Inoculate soil with microorganisms purchased from a garden center or cultured on-site.

5. Plant densely: about one plant every two square feet.

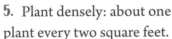

6. Cover with a thick mulch layer.

7. No management is the best management. Water and weed for the first few years as needed.

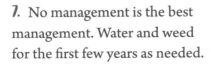

"IF I'D HAVE MOVED IN AND PLANTED A JUNGLE, THE NEIGHBORS WOULD'VE REPORTED ME TO THE PLANT POLICE. I HAD TO DO IT SLOWLY AND SNEAKILY."

THE HABITAT BUILDER

DON MAHONEY

Richmond, California

WHEN DON MAHONEY BOUGHT HIS BAY AREA HOME IN 1996, IT came with a vacant lot next door. "If I'd have moved in and planted a jungle, the neighbors would've reported me to the plant police," he said. "I had to do it slowly and sneakily."

As horticultural manager and then curator at the San Francisco Botanical Garden's Strybing Arboretum, Mahoney had plenty of ideas about what he'd like to sneak onto the property. There was already a walnut tree on the lot, along with a tangle of wild blackberries and ivy that he had to remove.

Then he started planting trees, as densely as he could. "I hadn't yet heard about the Miyawaki method, the Japanese approach to intensely planting a forest in a small space, but that's exactly what I was doing," he said.

He set about filling the land with unusual trees. He put in a monkey's hand tree (*Chiranthodendron pentadactylon*), a Mexican native, related to the hibiscus, that has brilliant red flowers with five finger-shaped stamens. He also planted a rare redwood with bluish foliage, similar in color to a blue spruce. He added a willow and a daisy tree, *Montanoa grandiflora,* which is not technically a tree at all, but a large woody shrub. "It gets up to tree size," he said. "That's the one that grows in Mexico where the monarch butterflies overwinter."

That was the purpose of his tree collection from the beginning: to attract insects, birds, and other wildlife. "Trees are the backbone of a

habitat," he said. "It's all about the biodiversity for me. If it's a common tree that's everywhere already, I won't plant it."

His inspiration came from the British hedgerows planted between fields and pastures. "They look wild and unkempt, but they actually get quite a bit of upkeep. You have to cut down a fifth of your hedgerow every five years. If you don't, it gets decrepit and old and less biodiverse. It loses its value for insects that like fresh growth."

That means he has to be ruthless about cutting back trees, no matter how much he loves them. "I'm always on the lookout for what has to get cut next. I'll walk around and say, 'Okay, this little section of trees is going to get thinned out this year.' But I'm seventy-four. I can't climb trees anymore. A tree company comes by twice a year and does it for me. It looks like an untamed jungle, but there's a lot of maintenance that goes into it."

Mahoney's approach won over the neighbors, who make a point of walking by to see what's living in the habitat he created. "It's all planted on the street side," he said. "Anyone's welcome to come take a look."

Since his retirement from the arboretum, he keeps a logbook where he records the changes in bird and insect populations over the years. "The willow tree alone produces loads of pollen," he said. "I've seen warblers, finches, and sparrows up there, feeding on the insects gathering that pollen. Last Christmas I counted twenty-two species of birds. I have an Oregon oak (*Quercus garryana*) where butterflies and moths lay eggs. I've spotted ten species of native bees. Every year it changes—all of these creatures go through cycles. I'm learning so much just by keeping track of what comes through here."

"TREES ARE A LONG-TERM PROPOSITION FROM SEED."

THE SEED COLLECTOR
DEAN SWIFT
Alamosa, Colorado

TREE SEEDS ARE UNIQUELY DIFFICULT TO COLLECT. THEY'RE only ripe and available for a short window of time, and that window shifts subtly from year to year. They can be difficult to reach, clinging stubbornly to the uppermost branches. And they're a prized food source for birds and small mammals, so there's plenty of competition.

Dean Swift began collecting seeds as a child, working in his parents' tree nursery near Denver. "My dad liked to explore the mountains and the countryside, so we would gather seeds, especially conifer cones."

At the age of twenty-five, he moved to "the middle of nowhere" in southern Colorado, where he collected the seeds of conifer trees and native wildflowers. Every year, he would send a one-page list of offerings to a select group of seed companies, tree nurseries, and Christmas tree farms. "It's a small world, basically," he said. "They all know me and I know them. My whole mailing list is about three hundred names."

That simple catalog mailing has a complex operation behind it. Swift specializes in fir, spruce, and pine seed from Arizona, New Mexico, Colorado, and South Dakota. He employs seasonal crews of seed collectors, all of whom he's trained himself. "It's kind of a family tradition," he said. "I've counted five generations out there, if you include the little kids."

White fir
(*Abies concolor*)

Not everyone on the crew is scrambling up a tree.

Ponderosa pine
(*Pinus ponderosa*)

"In Europe they climb the trees. We're quite fortunate in North America to have a little critter called the red squirrel. They go up and get the cones as soon as they're ready in September. They're infallible as far as seed maturity goes. I've seen them cut the cones off one tree and wait a week to cut cones off the tree beside it."

The squirrels hide caches of cones around the forest, and Swift and his crews go looking for them. (This doesn't rob the squirrels of a winter's food supply: they collect far more than their human competitors will ever find, and they have other food sources.)

Once the cones are gathered, Swift dries them in the sun. "In the Pacific Northwest they use natural gas or heating oil in big kilns to dry the cones. But here in Colorado, I just put them outdoors in burlap bags. It's my solar dryer." The dried cones are tumbled in a wire mesh screener to shake out the seeds, and then a gravity separator sorts the empty seeds from the viable ones.

The seeds have to be carefully labeled by species, but also by region. The small genetic variations within species, from one region to the next, are referred to as ecotypes. "The genetic ecotype is really important now," Swift said. "Take blue spruce. There's one we collect in northern Arizona that's a little slower growing, but it's also the most blue, and it starts growing about a week later in the spring than the others. So in areas that are subject to spring frost, this one works better. A good nursery knows the difference."

The subtle distinctions between ecotypes might help identify trees that are more resilient

Blue spruce
(*Picea pungens*)

in the face of climate change. "Trees are a long-term proposition from seed. You can't plant a tree based on what the conditions are like right now. You have to think about what the climate might look like in thirty years."

Tree seeds are a marvel of durability and longevity. Swift stores his inventory in a big freezer, where they remain viable for decades, serving as a kind of insurance policy against lean years, and as a retirement plan for the seed collector himself. "I just sold the last of a batch of pine

Douglas fir
(*Pseudotsuga menziesii*)

seeds I collected in 1991," he said. "They were still germinating at 93 percent. I don't mind getting in a ten- or fifteen-year supply when I can, because I don't know when I'll get another chance."

NON-HUMAN SEED COLLECTORS

CLARK'S NUTCRACKER

Collects the seeds of whitebark pine (*Pinus albicaulis*) and others, hiding up to one hundred thousand seeds in a single year for miles around the tree. The birds can return unerringly to every cache, but they don't eat them all. The seeds left behind are buried at the perfect depth for germination and are essential to the survival of the pine.

ACORN WOODPECKER

Drills small holes into dead trees with its beak and uses those holes as a granary to store acorns. A granary could have tens of thousands of holes and be managed collectively by many birds. As the acorns dry out and shrink, they must be relocated to smaller holes, requiring constant maintenance throughout the season.

CALIFORNIA SCRUB JAY

Hides pine tree seeds and other food in thousands of locations and moves them around constantly to protect against thieves. Scientists have observed that jays are more likely to move a cache if another bird watched them hide the food, and that jays who are good thieves themselves are more alert to the danger of robberies.

AGOUTI

Buries seeds of tropical South American trees and moves the cache a few dozen times to protect against predators. This rodent, the size of a house cat, has taken over seed dispersal from massive, long-extinct mammals who used to do the job, helping to save trees like the black palm (*Astrocaryum standleyanum*) from extinction.

SPIDER MONKEY

Feeds on the fruit of
tropical trees in South
American forests.
They tend to swal-
low fruit whole,
digesting the seeds and depositing them,
a few hours later, with a dose of fresh fertil-
izer on the forest floor. A single spider monkey
might disperse 195,000 seeds in a year.

RED SQUIRREL

Climbs conifer trees to clip their
cones, then stores them in an
aboveground pile, called a midden,
made up of fresh cones and the de-
bris left behind after the seeds are
eaten.

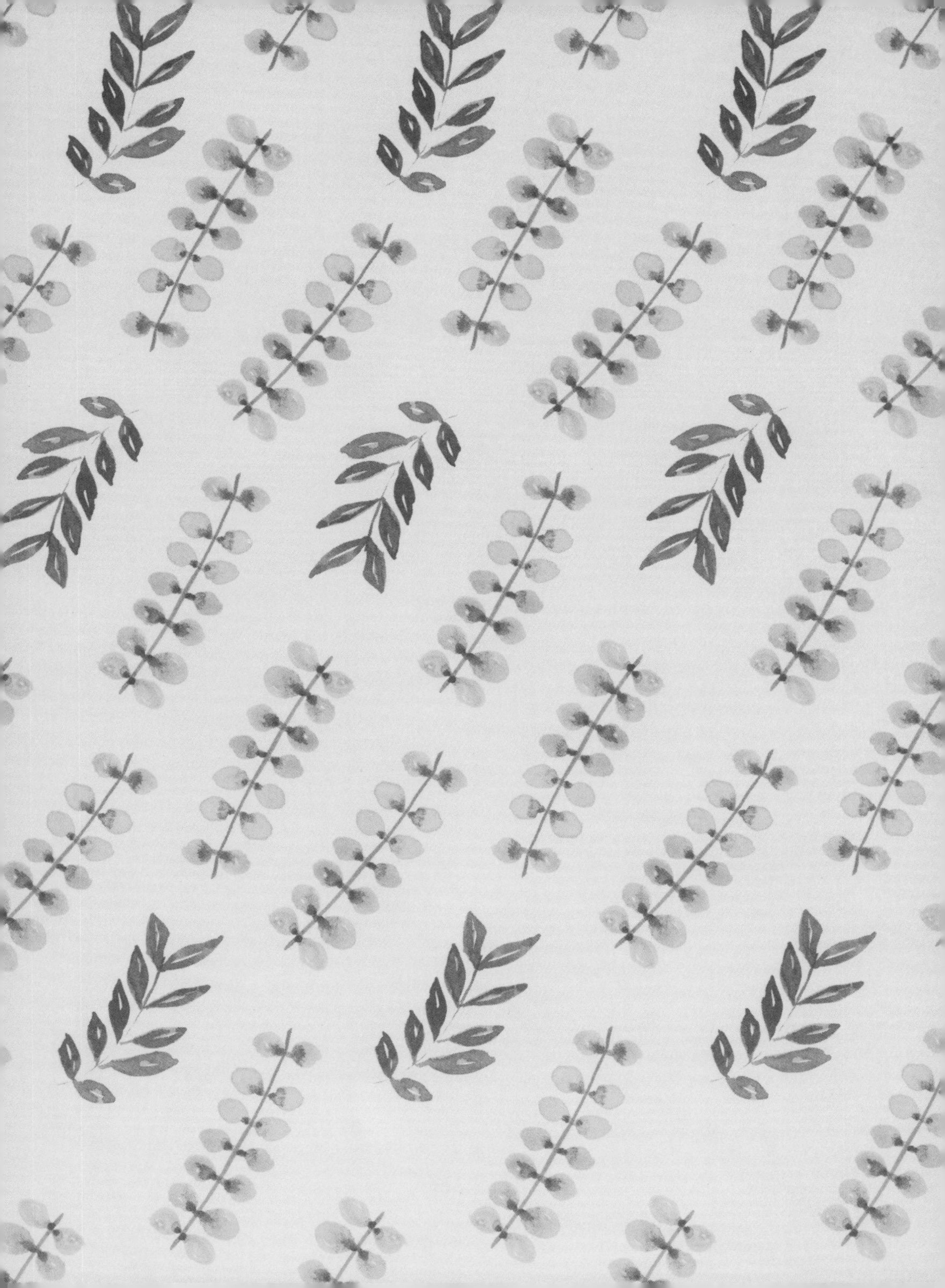

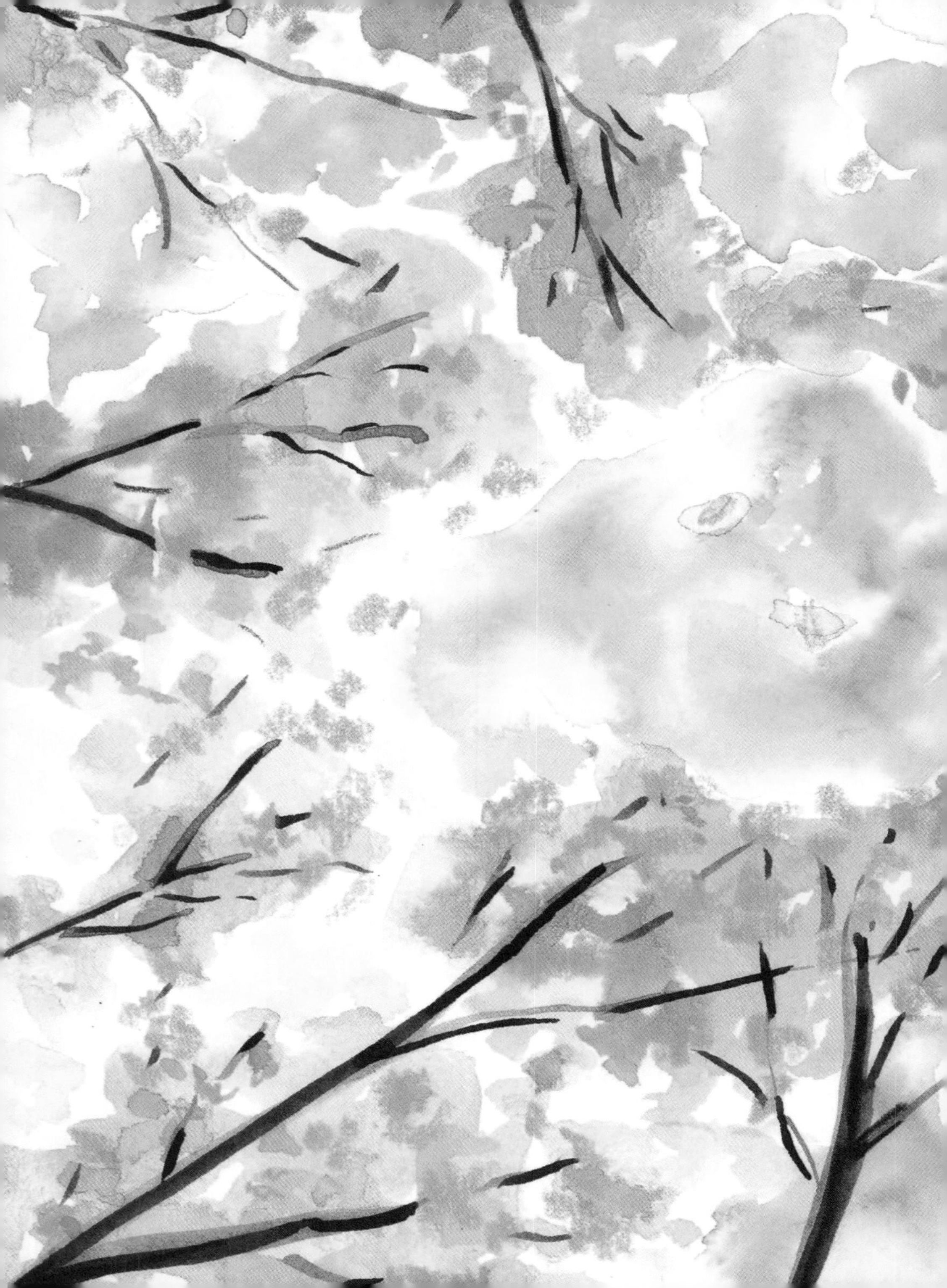

ARTISTS

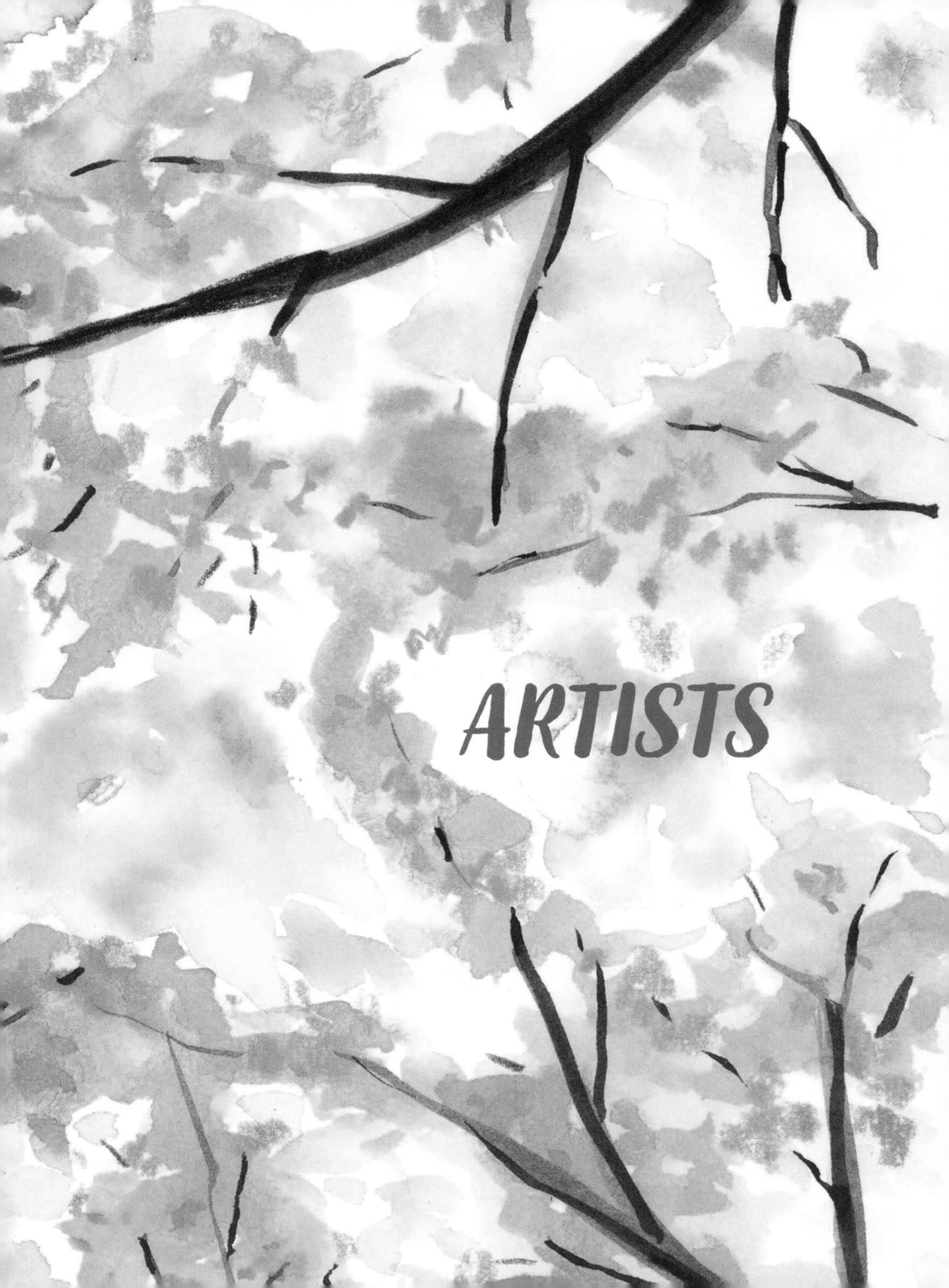

"IF I COULD DO THIS WITH HAIR, WHAT COULD I DO WITH A TREE?"

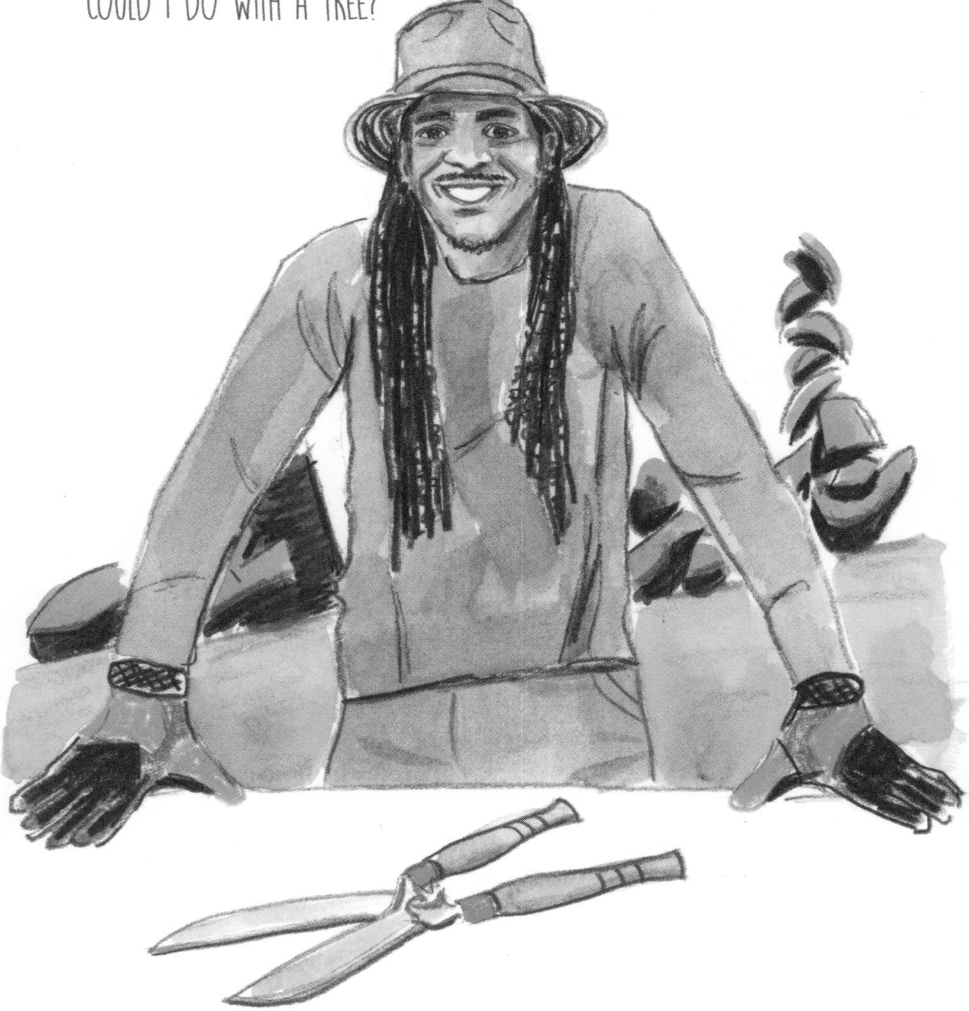

THE ABSTRACT ARTIST

MIKE GIBSON

Columbia, South Carolina

"WHAT REALLY GOT ME INTO TOPIARY WAS CUTTING HAIR," SAID Mike Gibson. "My mother was a beautician, and I learned how to trim and make designs when I was a kid. I would use a hard gel and sculpt my hair into different shapes. By high school I was doing it for all my friends. That got me thinking, *If I could do this with hair, what could I do with a tree?*"

After high school he studied art, trying charcoal, pen, acrylic, and sculpture. He also worked as a landscaper, trimming trees and bushes for his neighbors in Youngstown, Ohio. He started asking his clients if he could make an abstract shape in their landscape. "I'd see a tree and think, *I can do something with that.* So, I'd make a drawing for the client, and sometimes they'd practically dare me to try it. And you know what, I found out I could do it. If I could draw it, I could put it into a tree."

He called himself a property artist. "I didn't use the word *topiary*. That was pom-poms and spirals. Nobody else was doing property art, so I thought that could be my own thing."

Then he went home and showed his father, who is also an artist, what he'd been working on. "Hey, look at my property art," he said to his dad. "Nobody's doing this." But somebody else was doing it, and Gibson's father knew all about him. "My dad told me about Pearl Fryar," Gibson said, "and that changed my life."

Pearl Fryar began sculpting abstract topiaries in his garden in Bishopville, South Carolina, in the 1980s. Since then, he's gained notoriety in horticultural and art circles for his fantastic creations, which more closely resemble a Picasso or a Henry Moore sculpture than the clipped boxwoods seen at amusement parks and shopping centers.

"I heard about Pearl Fryar on a Saturday," Gibson said. "That began a week of discovery. I went to the library, I watched every video, read every article, I had to learn everything I could about this man. Then I started to wonder if there were more Pearl Fryars. But I couldn't find anyone who was like him. Pearl, to my knowledge, was the only African American doing this. So if he was the only one, I needed to be the second one." Over the next several years, he visited Fryar often, learning everything he could. "Pearl just poured his wisdom into me," he said.

He knew that Fryar's topiaries had inspired beautification efforts around Bishopville, and he saw an opportunity to do something similar in Youngstown. "It became a mission. I wanted to help change the narrative about Youngstown. I wanted people to visit from all over the world and say, 'Hey, this is the most beautiful place. I've never seen this much topiary.'" He joined local revitalization efforts and put topiaries into the landscapes of newly restored homes and businesses. "Our area code is 330, so I was on a mission to put 330 topiaries around the city. It took me six years, but I did it."

Fryar, now in his eighties, has been unable to maintain his own creations lately. Gibson returned to Bishopville to help restore the garden, and he's been commissioned to tend Fryar's topiary installations at museums. "I try to bring my own style, but still pay homage to Pearl. You can't paint exactly like Monet, but you can retrace it, and add your own subtle variations to what he did."

He now lives with his family in Columbia, South Carolina, not far from Fryar's home. He's working in the University of South Carolina's African American studies program, organizing workshops in the arts.

His topiary work continues, even at the house his family rents. "I did, of course, sculpt the bushes in front of our house. I couldn't let them go untouched. And I have some potted topiaries I've been working on. And I did the neighbors' yards. I had to."

Most of his other work is commissioned by museums and universities. His style is recognizable for the repeating patterns he puts into trees and shrubs. "I'm inspired by the Fibonacci sequence, and the connectedness of triangles," he said. "At any given angle of any piece that I do, you'll be able to see a triangle. There's also a subtle spiral that you might not even notice, but it's a capital G, for Gibson. It might be clockwise or counterclockwise, but it's there. That's my signature. I'm leaving my mark. I hope others will follow and make their own mark."

"I THOUGHT GRAFTING WAS THE PERFECT METAPHOR FOR CONTEMPORARY EXISTENCE. IN SO MANY WAYS, I FEEL LIKE OUR LIVES ARE ALL SO PIECEMEAL AND HYBRIDIZED AND PATCHED TOGETHER."

THE CONCEPTUAL ARTIST

SAM VAN AKEN

Syracuse, New York

DOES A SINGLE TREE COUNT AS A TREE COLLECTION? IT DOES IF it's a Tree of 40 Fruit.

As an art school student, Sam Van Aken started experimenting with a concept borrowed from orchardists. "I was interested in the way they could graft one variety of fruit tree onto another," he said. "So I tried to do some version of that as conceptual art. I was taking pieces of plastic fruit and sticking them together. I thought grafting was the perfect metaphor for contemporary existence. In so many ways, I feel like our lives are all so piecemeal and hybridized and patched together."

His interest in plastic fruit didn't last long. He wanted to attempt the real thing. "My great-grandfather made a living working in orchards and grafting. I never met him, but everybody talked about him with this great reverence, because he knew how to graft. I wanted to learn how to do what he did."

He explored heirloom fruit varieties and was surprised to discover their unusual flavors. "For instance, these old plums are full of spice. I started to understand wine better through fruit, because you taste some things in the front of your palate and other flavors on the finish."

The history of those old varieties surprised him too. Because the seeds are not genetically identical to the parents, each variety can only be reproduced through grafting. "I have some varieties that are thousands of years old. That means that at least every fifty years, somebody's taken a branch from one of those trees and grafted it onto another tree. You see yourself as a tiny blip in its existence."

Once he had a handle on how grafting worked, he came up with an art project that he called the Tree of 40 Fruit. He would graft forty different stone fruit varieties onto one plum tree to create a spectacular color scheme when the tree bloomed.

Starting in 2011, Van Aken began planting his grafted trees on university campuses, outside art museums, and in public parks and gardens. The process wasn't as simple as grafting forty branches onto a single tree and planting it. The varieties had to be suitable for the climate where they'd grow, and each new graft had to be compatible with the branch it was being grafted onto. The process could take years.

"I have to use a ton of interstock varieties," he said. "For instance, cherries and plums don't readily graft together. I had to find a variety that I could graft between them to make it work." A Spanish plum called Puente, or Adara, will graft with apricots, almonds, plums, cherries, peaches, and nectarines, allowing it to serve as a sort of bridge between the host tree and the graft.

Each graft has to be made at a particular point in the growing season, and then Van Aken waits for the new growth to become established before adding the next round. "I'm usually coming back for three to five years to finish the process," he said. "There's trees I planted twelve years ago that I still visit."

The trees are established at the Everson Museum of Art in Syracuse and at Syracuse University, where he is a faculty member. The Children's Discovery Museum in San Jose, California, hosts a tree, and one has been planted at a hotel in Bentonville, Arkansas. His latest project is the Open Orchard on Governors Island, New York, a public orchard of 102 grafted trees.

Along with each grafted tree, he creates a kind of horticultural archive. He researches each variety's history going back several hundred years and makes botanical drawings. He presses the trees' dried leaves and flowers to make his own herbarium specimens. The result is very much like a botanist's notebook. "The idea to create a Tree of 40 Fruit was tied to this index/archive relationship," he said. "The tree is the index to this archive I've made that's historical and geographic."

In keeping with his original vision of a tree that produces a colorful palette of blooms, the mature trees are, in fact, stunning when the blossoms open in spring. But beyond the aesthetics, the history, and the science, there is the fruit itself.

"The longest season for any of these trees would probably be cherries as early as June, and peach and plum varieties going late into September," he said. "On a practical level, it's perfect for a home garden. You're getting fifteen to twenty pieces of fruit a week, as opposed to a few hundred plums ripening all at once and you have to figure out what to do with them."

Although his trees are more often planted in public spaces as works of art, they also feed a community. "I found out that people who live near the trees will put the harvest dates on their calendars," he said. "They'll show up with bags and collect the fruit."

TYPES OF GRAFTS

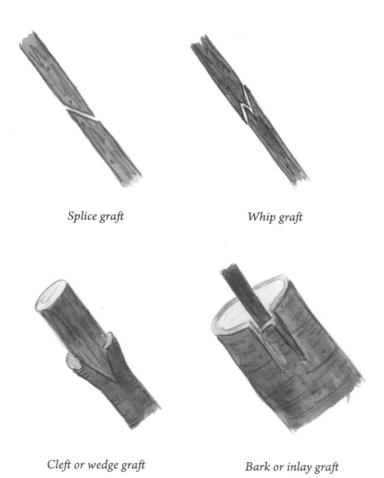

Splice graft

Whip graft

Cleft or wedge graft

Bark or inlay graft

"IF YOU KNOW HOW TO READ THE STORY, THE TREE TELLS YOU EVERYTHING."

THE BONSAI ARTIST

ENRIQUE CASTAÑO

Mérida, Mexico

"You don't see many second-generation bonsai collectors," said Enrique Castaño, a professor of biochemistry at a botanical research center in Mexico's Yucatán Peninsula. "Kids can develop a bad feeling about their father's bonsai trees. Either they've been told so many times to be careful around the trees that they don't want to go near them, or they resent that their father spent more time with the trees than he did with them."

Both his father and grandfather were sculptors. His father became interested in Japanese art, which led him to the art of bonsai. It's easy to see why bonsai would appeal to a sculptor: a bonsai is judged not only by what is there, but also by what isn't there. "The open spaces between the branches matter, sometimes more than the positive spaces, because they create the illusion of a large tree," Castaño said. In addition to the handful of bonsai trees created by his father, he lives among some of his grandfather's sculptures on eight acres in Mérida, Mexico.

"When I bought this land, it had been burned to the ground. That's how they clear land here. The thinking is that a property belongs to someone once it is burned. It is a mark that they own this land."

Over the last couple of decades, he's filled his land with an assortment of mostly native trees that can survive Mérida's hot and humid climate. At the center of his collection are the bonsai, arranged on tables near his house where he can keep an eye on them. "They require daily care," he said. "It's not like having a dog or a baby. A dog will bark, a baby will cry. You have to pay attention and involve yourself with the trees. This creates a connection of care that allows you to focus on something besides yourself. This is part of the healing power of bonsai."

Bonsai pots are small and shallow to constrict the roots and slow the growth of the tree, but that means they dry out easily. If he wants to go on vacation, he first undertakes an elaborate process of sinking each pot into the ground, so that they'll be easier for a gardener to water and won't dry out as quickly. He doesn't even dare to guess how many bonsai trees he's caring for: "A thousand. Ten thousand. I have no clue."

As the founder of a tropical bonsai society in Mexico, the author of a book on the botany of bonsai, and the president of Latin America's bonsai society, he's been able to lead local bonsai artists on plant collecting trips in Mexico, and to travel around the world teaching and exhibiting his trees.

Some of these trees tell the story of where he lives. In 2009, he exhibited a bonsai buttonwood tree, *Conocarpus erectus*, which is adapted to strong hurricanes and tropical storms along the coast. This particular species is known for its ability to grow new leaves on old, gnarled wood. He picked up a piece of that so-called deadwood along the coast, in an area that had been cleared of trees to make way for a new road. He kept it in the ground for three years, watering it and waiting to see if it would put down roots and form new growth. When he exhibited it, he wrote, "It has taken ten years to bring this tree from an almost dead piece of wood to a bonsai."

The great pleasure in tending bonsai comes from learning how to read the tree by watching the size of the leaves, the curvature of the branches, and the direction of new growth. "The trees can tell you what's going on right now and what was happening weeks ago, months ago, or years ago. Everything is written there, from the oldest part at the base of the trunk to the current history at the point of the leaves. If you know how to read the story, the tree tells you everything."

BONSAI SIZE CLASSIFICATIONS

Bonsai trees are categorized according to their size, with some artists specializing in larger or smaller trees. They are often described as "one-handed," "two-handed," or even "eight-handed," depending on how many hands are required to lift the tree.

KESHITSUBO
seedling size, 1 to 3 inches

SHITO
fingertip size, 2 to 4 inches

MAME
palm size, 2 to 6 inches

SHOHIN
one-handed, 5 to 8 inches

KOMONO
one-handed, 6 to 10 inches

KATADE-MOCHI
one-handed, 10 to 18 inches

CHUMONO *or* CHIU
two-handed, 16 to 36 inches

OMONO *or* DAI
four-handed, 30 to 48 inches

HACHI-UYE
six-handed, 40 to 60 inches

IMPERIAL
eight-handed, 60 to 80 inches

"I WANT THEM TO
SEE WHAT I SEE."

THE CHRONICLER OF GINKGOES
JIANMING (JIMMY) SHEN
Hangzhou, China

Hangzhou, a city of twelve million in eastern China, is known for its cultural riches and natural beauty. It has been continuously inhabited for more than two thousand years and straddles the ancient Grand Canal, the longest of its kind in the world. At the western edge of the city is Xi Lake, also called West Lake, one of the most tranquil bits of scenery in China. Surrounding the lake are a botanical garden, a zoo, an art museum, and a breathtaking hilltop pagoda offering city and countryside views in every direction.

Photographer Jimmy Shen lives in a suburb of this beautiful city but finds his inspiration in the mountains about fifty miles to the west. The Tianmu Mountain National Nature Reserve is the only place in the world where a naturally occurring population of wild ginkgo trees might still be found. There is no way to prove definitively that the enormous old trees growing there today weren't cultivated by humans. Ginkgoes were almost certainly planted a thousand years ago by Buddhist monks at a temple near the top of the mountain, and it would be impossible to distinguish those trees from truly wild ones.

Shen worked in advertising and was often hired to photograph buildings and scenery. But throughout his career—going back to the days before digital photography, when all his images were on slides—he has been photographing ginkgo trees in the region.

His aim was to document these ancient ginkgoes and to help people appreciate them. "In China, because we have a lot of ginkgo trees, they are maybe not so interesting," he said. "People see them with their own eyes. I want them to see what I see."

The ginkgo is a living fossil, which is to say that it has continued to flourish, largely unchanged, since its appearance in the fossil record two hundred million years ago. It is the only living relation of an entire subclass of trees that once grew all around the world, alongside the dinosaurs. Its relatives slowly died off, and by the end of the Pliocene epoch, about two and a half million years ago, *Ginkgo biloba* was the only member of its family left. It nearly disappeared too, vanishing from the fossil record everywhere except this region in eastern China where Shen lives.

With their unique fan-shaped leaves that turn to a dazzling gold in fall, ginkgo trees appealed to seventeenth-century European botanists and explorers, who collected their seeds in Japan, Korea, and China. As a result, the tree is now widespread across Europe and North America, as well as Asia, and it appears sporadically in the southern hemisphere as well.

"Ours are different from the ginkgo trees you have in the United States," Shen said. "Ours are much more vibrant. We have a lot of varieties that you don't have. Yours are mostly imported from Japan. There is not so much variety. And you mostly plant the male. We have the females, because we Chinese eat the nuts." Ginkgo nuts are toxic, but in China they are cooked and eaten in small quantities as a medicinal plant or health food.

While the trees are interwoven into Chinese culture, Shen felt that there were particular trees—ancient trees, historically significant trees—that people tended to overlook. It would be impossible to collect the trees themselves—they have to remain where they're planted—but eventually he realized that he could collect images and stories about them. He reached out to botanists, archaeologists, paleobiologists, and other experts within China and around the world to help him. The result is *The Wild Ginkgos,* a book of photographs and descriptions of significant trees.

He cataloged four-thousand-year-old ginkgoes that stand over a hundred feet tall. He documented a two-thousand-year-old ginkgo that was struck by lightning in the distant past, leaving a hole inside the trunk "that can accommodate thirty people." He wrote of ginkgo couples, celebrated male and female trees that have stood side by side for centuries and now preside over weddings beneath their branches. The couples

choose these sites because they hope their marriages will enjoy the same longevity that the trees have.

In very rare cases, a ginkgo will sprout both male and female branches on one tree, sometimes with the male branches on one side and female on the other, and sometimes with the male branches up top and females below. He was able to document seven of these monoecious ginkgoes, the youngest of which was four hundred years old.

Other wonders include trees that produce unusual sounds as the wind whistles through them, an eighteen-hundred-year-old tree with such an elaborate ground-level root system that a hundred chickens roost inside, and the "Most Miserable Ginkgo," which has been cut down six times in its thousand-year life and has regrown every time. Today ten trees sprout from its long-suffering stump.

These stories offer a window into the role that ginkgo trees play in Chinese culture and everyday life. Shen sees the trees as China's goodwill ambassador. "Most people don't know about the history of the ginkgo. My book is only a peephole, a small view into the ginkgo trees of China."

LIVING FOSSIL • An organism, such as a plant or animal, that lives today in a form more or less unchanged from the fossil record.

LAZARUS TAXON • A species that was thought to be extinct and seemed to have come back from the dead, referring to the biblical story in which Jesus raised Lazarus from the dead.

ZOMBIE TAXON • Fossils from a much older era that have been deposited into more recent fossils, making it seem like they were alive in an era when they should have been long dead.

ELVIS TAXON • A species that is similar to an extinct creature; a look-alike or impersonator.

GINKGO OR MAIDENHAIR TREE
GINKGO BILOBA

Fossil records date back two hundred million years. All close relatives are extinct. Has

been cultivated in China for more than a thousand years and is now widespread in Europe and North America as well.

DAWN REDWOOD
METASEQUOIA GLYPTOSTROBOIDES

Believed to be extinct until 1943, when Chinese botanist Kan Duo spotted it growing in Hubei province. Seeds were collected and distributed to botanical gardens worldwide. It is one of only three redwood species in the world, the other two being the giant sequoia (*Sequoiadendron giganteum*) and the coast redwood (*Sequoia sempervirens*).

MONKEY PUZZLE
ARAUCARIA ARAUCANA

Native to Chile and Argentina. Fossil records date back two hundred million years. Thick, reptilian leaves made the tree an interesting novelty to European explorers, who introduced it to botanical gardens. Today it grows throughout North and South America, Asia, Europe, and Australia.

WOLLEMI PINE
WOLLEMIA NOBILIS

In 1994, a group of Australian botanists found about two hundred of these trees growing in a remote canyon in a national park near Sydney. Seeds have been carefully collected so that the tree can be grown in gardens around the world.

"A TREE BECAME A KIND OF SYMBOL OF BELONGING, OF BEING ROOTED SOMEWHERE, OF STABILITY, OF STANDING—STANDING ON SOMETHING, OR STANDING FOR SOMETHING."

THE FILMMAKER
SALOMÉ JASHI
Tbilisi, Georgia

In 2015, a photograph of a tree floating across the Black Sea went viral in the nation of Georgia. It was an unsettling sight: What was a full-sized tree, with its high and graceful canopy, doing onboard a barge? Where was it going? And most important: Why?

Those questions were answered almost immediately, with news reports that former prime minister and billionaire Bidzina Ivanishvili was buying mature trees, often from rural and impoverished Georgians, and having them transported, at great expense, to a park he was developing in the coastal town of Shekvetili. The event sparked protests over the environmental impact of uprooting the trees.

When Georgian filmmaker Salomé Jashi saw the photograph, she shared the protestors' concern over the decision to buy and transplant mature trees. But she saw something else as well. "I was mesmerized at the sight of this floating tree," she said. "It felt like an error, like a glitch. The ambivalence of this image being beautiful and ugly at the same time attracted me straightaway. We went to film the very next day."

The result is the documentary *Taming the Garden,* which was released in 2021. The film is a meditative, almost poetic depiction of enormous trees being painstakingly dug out and transported away from their homes. The people who agreed to sell their trees to Ivanishvili are on camera as well, sometimes quietly bickering with spouses over the decision, sometimes crying as the tree rolls away. There's no narrator to explain the situation to the viewer. The camera acts only as a witness to the strange spectacle unfolding before it.

It wasn't easy for Jashi to find families who were willing to allow the process to be filmed. "We live in a very fragile democracy, where local governments tend to be quite controlling, and it's mostly public institutions that employ people. They don't want to get into trouble." She recalled one woman who didn't want to sell an enormous old tree that still held carvings made by her parents when they were young. Her husband insisted, and in the course of the tree being excavated, it fell and broke. The woman said that it was as if a member of the family had died. "When she told me, she was crying and I was crying," Jashi said. "But she did not want to be on camera."

Some people were happy to sell their trees. "In some cases, the tree meant nothing. It crowded the house. It shaded the garden. Some families were happy to get rid of a tree, and to have the money." There were also some community benefits, as roads were built or expanded to accommodate the enormous trucks hauling the trees away.

No matter how the people in the film felt about selling their trees to a billionaire oligarch, the viewer can't help but have a reaction to it. There's something horrifying about watching a grand old tree being dug up and carted off. But in the next moment, a viewer might think, *We all do this. We all participate, as consumers, in the cutting down of trees. What's the difference?*

Jashi agrees. "It's a matter of consumerism. There's also an element of colonialism, in the sense of these trees being taken, sometimes in return for what you might call civilization: money for education, or money for cancer treatment, or new roads. As privileged people, we can be critical of that, but for these people, it was a way to save their lives, or improve their lives. Maybe it is a kind of Faustian deal."

Jashi estimates that two hundred trees were moved at a cost of about $300,000 each, including oaks, lindens, magnolias, and chestnuts. She finds the Shekvetili Dendrological Park, which opened in 2020, to be a strange experience for visitors. "There's no proper design," she said. "It's like a parking lot, with trees in rows. And they did not all survive. Some died as they were being transported, and some died after they were planted." Local news accounts about the park describe people coming to visit and trying to find the tree they sold to Ivanishvili, but some of them had been pruned so severely that they were unrecognizable, even to those who lived under their branches for decades.

As she worked on the film, Jashi came to see the tree as a symbol of something else. "When I first saw a tree moving, I felt dizzy. You know when you're on a roller coaster, and your axis shifts? You don't know where is the ground, and where is the sky? I had this feeling that my axis had shifted. And for me, a tree became a kind of symbol of belonging, of being rooted somewhere, of stability, of standing—standing on something, or standing for something."

HOW TO MOVE A LARGE TREE

There are many worthy reasons to move a tree: to save it from a construction project, to place it in better growing conditions, or to allow more space for other trees to grow. Regardless of the motive, the technology behind moving large trees is quite sophisticated. Here's how it works:

BALL AND BURLAP • Dig a trench around the root ball with shovels or an excavator. Dig under the root ball as well, until the tree sits on a small pedestal of earth. Wrap the root ball in burlap and rope and pull the tree off the pedestal using a crane or forklift.

TREE SPADE • Use a truck-mounted tree spade, a motorized assembly of enormous steel blades that surround the tree, to dig into the soil and pull the tree out. The tree can remain on the truck, the root ball held together by the closed blades, and the machine will plant it directly in its new location.

STEEL PIPES • The most enormous trees are moved by digging a trench around the edge of the canopy, driving steel pipes underneath the root ball to form a sort of metal platform, and then using hydraulic equipment to lift the tree onto a truck for transporting.

SURPRISING FACTS ABOUT MOVING TREES

From Dave Dexter, founder of Dexter Estate Landscapes and father of Lucas Dexter (see page 227):

YOU DON'T NEED TO DIG VERY DEEP. It differs by species, but the active root zone of most trees is only about eighteen inches deep.

IT'S BEST NOT TO PRUNE A TREE BEFORE MOVING IT. Rooting hormones are generated in the tips of branches. Cutting the tree back makes it harder to put down roots in a new location.

TREES DON'T NEED RICH SOIL WHEN THEY MOVE. A little compost is nice, but a tree wants to be planted in familiar native soil, not a rich mixture.

MOVING A TREE IS A SHORT-TERM INVESTMENT. If you want a big tree now, you'll pay to move it. But over the years, a smaller, nursery-grown tree will catch up in size to a larger tree that was transplanted at the same time.

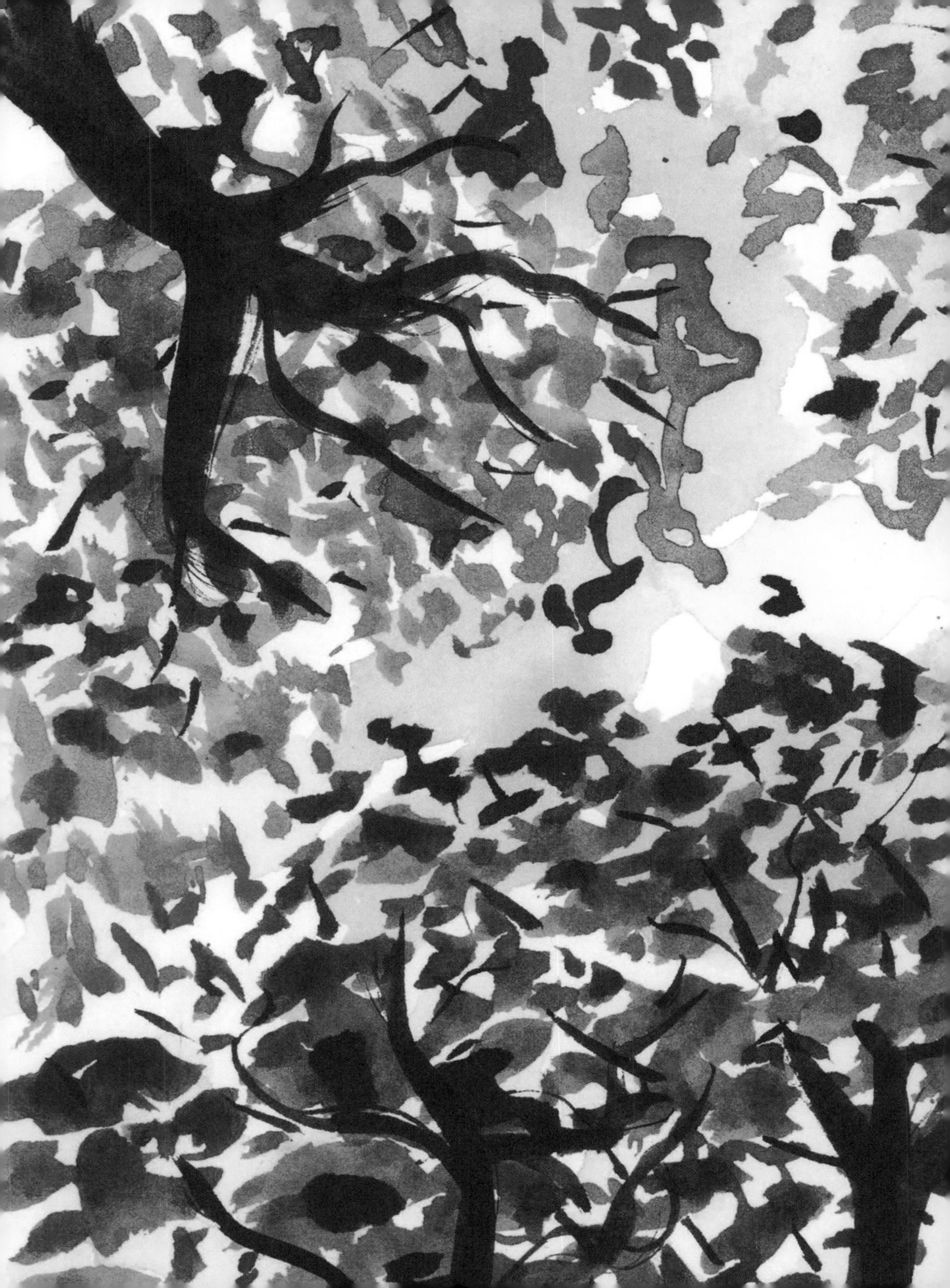

CURATORS

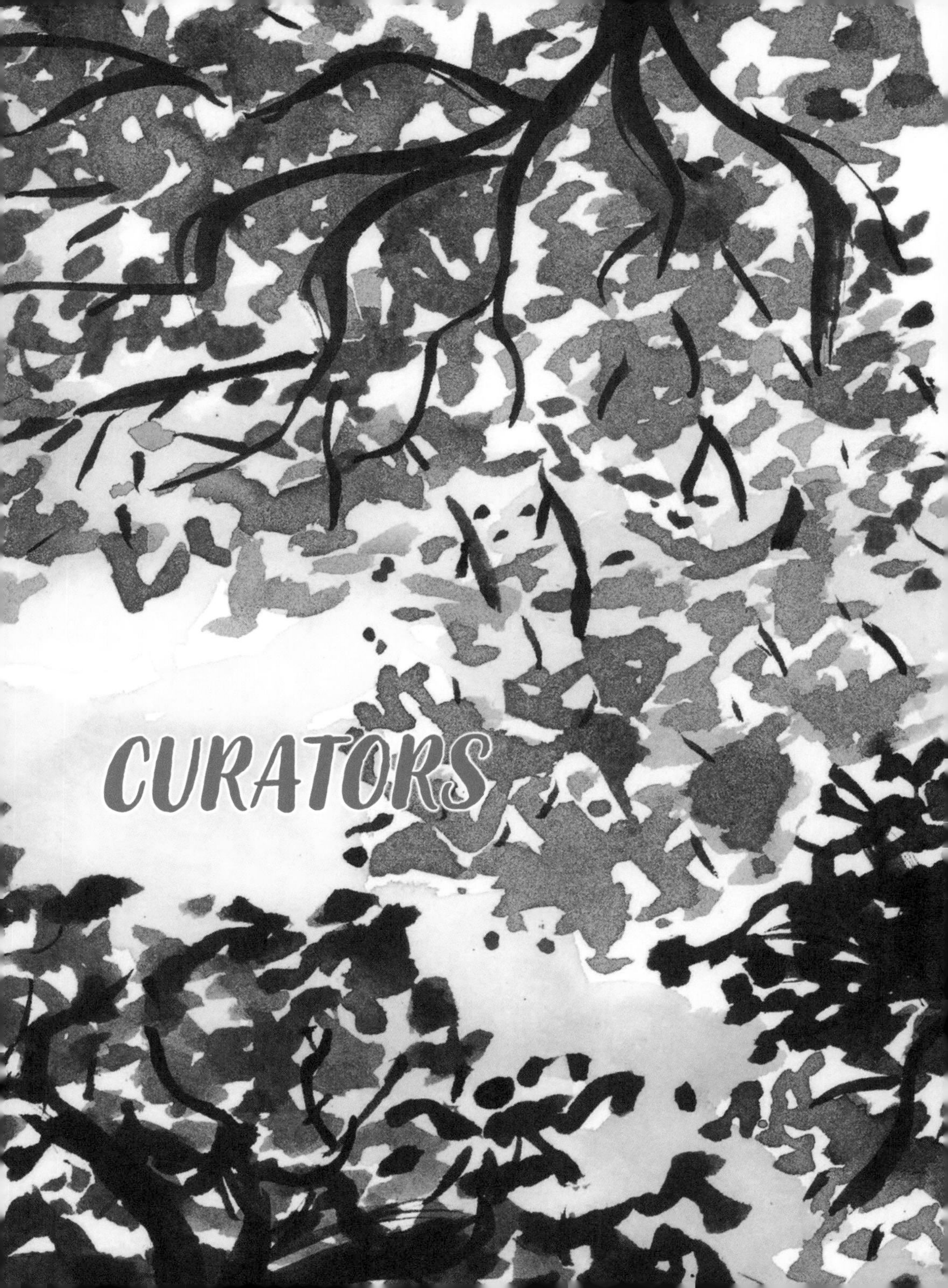

"MAYBE IT'S FUTILE, BUT
I ONLY WISH THAT EVERYONE
COULD HAVE SUCH A FUTILE
ACTIVITY."

THE OAK COLLECTOR

BÉATRICE CHASSÉ

Saint-Jory-de-Chalais, France

It took another collector to inspire Béatrice Chassé to collect trees. After a career in business communications in Brussels, she wanted to move to France and had an idea that she might start a botanical garden there. "My companion, all his life he had collected things—stamps, china, and other things that people collect. He told me, 'That's a good idea, to have a botanic garden, but you need to have a collection of some kind.' I said, 'What do you mean, a collection?' and he said, 'Well, I don't know, pick a tree.'"

She chose oaks. It wasn't a completely arbitrary choice: as she was looking for land to buy, it was obvious that *Quercus robur,* the common oak or European oak, was everywhere. "But it wasn't until after I'd settled on oak that I discovered that *Quercus* is one of the largest genuses of trees in the world. This was going to take some time."

When she started to learn more about oaks, their place in the natural world, and their ecological role, her reason for collecting them changed. "I wasn't just trying to grow every single oak. The collection became the means of understanding what this group of trees was about."

The land she bought—sixty acres in France's Dordogne region, bounded on one side by the river Côle—had been used for grazing cattle, but never for intensive agriculture. It was too hilly for that, and too rocky. But none of that would be a problem for the oak trees. "We have a lot of slopes that shelter trees from wind and cold and let the soil drain," she said, "which is good for planting trees that come from all over the world."

She began buying trees from specialty nurseries but quickly exhausted the supply that was available commercially. "Then I realized, 'Wow, what a fantastic thing, now I have to travel around the world to find these oaks.' After I understood the importance of provenance, I needed to know where the seed came from before planting a tree."

For first-time visitors to her arboretum, the real surprise comes when they see a tree like *Quercus viminea,* from northwestern Mexico, with long, skinny leaves that might, to the uninitiated, more closely resemble the leaves of an oleander shrub. "It provokes the most violent reaction from people. They insist it's not an oak tree." *Quercus salicina,* found in South Korea, Japan, and Taiwan, is a diminutive oak with long, glossy, deep-green leaves that looks nothing like a European or American oak. "After a while, people ask, 'How is it possible that all of these plants that don't look anything alike are all oaks? What makes an oak an oak?' And my reply is always that they know the answer if they just think about it for a second."

The answer, of course, is the acorn. An acorn is instantly recognizable and unique: it is large, easy to spot, packed with protein, and rich in fat, making it a valuable food source for any number of birds, rodents, and other mammals. It's also tricky to collect in the wild. You need to know when the acorns are ripe and ready to drop, and then plant them quickly, while they're fresh. While some seeds can germinate after decades or even centuries in storage, acorns lose their viability when they dry out. Even under perfect conditions, in cold, damp storage, they'll last only a few months if they're not put in the ground.

This makes it all the more remarkable that Chassé has been able to fill her arboretum with oaks grown from seed. Of the currently recognized 430 species, dozens of subspecies and varieties, plus countless numbers of hybrids, her collection includes 319 different oaks.

Does she stand a chance of collecting them all? "The species in Indonesia won't grow here. The ones that live in low altitudes in Asia, Central America, and southern Mexico will be difficult. Many of them have never been in cultivation. If they are, they're in three or four gardens around the world. And if you want to have a serious oak collection, you cannot collect acorns from a garden, because there's a very good chance they'll be hybridized with other oaks. In the wild, you go looking for a good group of the same species, so there's a much smaller chance that they've hybridized."

In 2012, Chassé's arboretum, called Arboretum des Pouyouleix, was certified through the French organization Conservatoire des Collections Végétales Spécialisées, which recognizes exceptional plant collections. The arboretum is open to the public from May to October, by appointment. It represents the largest oak collection in France and one

of the largest anywhere, with the trees arranged geographically according to every region in the world where oaks grow.

Like many tree collectors, she's aware that her collection will outlive her, and she doesn't know what will happen to it when she's gone. That doesn't stop her. "I know a person who is ninety-eight years old and he's been collecting oaks for forty or fifty years," she said. "I know another collector who is ninety-two, and one who is eighty-eight. They seem to be in such great shape. If this is the only reason for collecting oaks, that's good enough. Collecting is a passion, and that's what keeps people alive. Maybe it's futile, but I only wish that everyone could have such a futile activity."

PRIVATE TREE COLLECTIONS YOU CAN VISIT

More than seventeen hundred botanical gardens and arboreta welcome the public every year. Most are well-known, established gardens like the Royal Botanic Gardens at Kew in London or Longwood Gardens in Pennsylvania. Many of those institutions got their start as private collections a century or two (or three) ago.

But what about today's private collections? Some of them are open to the public too.

ARBORETUM WESPELAAR in Wespelaar, Belgium, is the collection of businessman Philippe de Spoelberch, whose family has been involved in Belgian brewing for centuries. While his business revolves around the global beer conglomerate InBev, he's also a noted tree collector who has dedicated a portion of his family's estate to a public arboretum. He relishes the challenge of editing his collection, creating vistas, and removing trees that simply fail to thrive. "You can buy any stupid painting and it will do nothing but sit there and go up in value," he said. "But a tree collector buys himself a problem."

ENEA TREE MUSEUM in Rapperswil-Jona, Switzerland, is the work of landscape architect Enzo Enea. He's created a tree museum, populated with trees rescued from construction sites and other areas where they might have been destroyed. Contemporary sculpture is on display

throughout the landscape, and many trees are placed against tall blocks of sandstone to allow visitors to better see the structure of the trees and view them as works of art.

GLANUSK ESTATE in South Wales is run by Harry Legge-Bourke, continuing the legacy of his father, William Nigel Henry Legge-Bourke, who built an extraordinary oak tree collection. Today the quercetum

consists of three hundred different oak species and cultivars on thirty-five hundred acres, which are only a small part of a much larger estate totaling twenty thousand acres. Of his father's interest in collecting oaks, Harry has said, "Start looking into oaks, and you soon see that there's only one sensible course of action: collect the lot."

HINEWAI RESERVE on Banks Peninsula in New Zealand is a remarkable three-thousand-acre nature reserve that can only be called a "collection" in the sense that nature has done the collecting. In 1986, an ecologically minded businessman named Maurice White met botanist Hugh Wilson, and a partnership was born. White bought the land, and Wilson has lived on it and managed it for the last thirty years. His unconventional idea—that invasive gorse should be left alone to provide a canopy for the forest to regenerate—seemed shocking at the time, but the decades have proven him right. Wilson's life and work on the reserve are unencumbered by the internet, cell phones, or a motorized vehicle. He walks or bicycles around the land and produces a handwritten newsletter twice a year, illustrated with his own drawings, which his more plugged-in helpers distribute to the reserve's supporters.

LOTUSLAND in Santa Barbara, California, is a funhouse of bizarre tropical whimsy created by Polish actress Madame Ganna Walska, who purchased the thirty-seven-acre estate in 1941. She designed a truly fantastical land-scape complete with topiaries, palms, cycads, and anything exotic and dramatic enough to capture her imagination. She died in 1984, but her garden, located in a secluded neighborhood, is open to visitors by reservation.

MEREWEATHER ESTATE in Victoria, Aus-tralia, represents fifty years of tree col-lecting on the part of Bill and Heather Funk. With more than two hundred spe-cies of oaks and an impressive conifer collection, the arboretum is set among fourteen hundred acres that include land for sheep grazing and guest cottages for rent. Heather remembers the days, long before the internet, when Bill would sit up and write a letter every night to a fellow collector or nursery owner, asking for seeds. "Little packets of seed would arrive in the mail," she said. "We were just starting out and we had nothing to spare, but there were always a few dollars here and there for trees."

NONG NOOCH TROPICAL GARDEN in Chonburi Province, Thailand, got its start in 1954, when Nongnooch Tansacha and her husband, Pisit Tansacha, bought six hundred acres of land with the idea of turning it into a fruit plantation. She spent the next twenty years collecting tropical trees and flowers. Today the garden, which is run by her son, Kampon Tansacha, contains a world-class collection of cycads, a dinosaur garden in which enormous replicas of dinosaurs are surrounded by palm trees and topiaries, a replica of Stonehenge, and other bizarre botanical displays, alongside event venues and resort-style lodgings.

POLLY HILL ARBORETUM on Martha's Vineyard is a twenty-acre collection of magnolias, camellias, stewartias, and anything else that caught the eye of horticulturist Polly Hill, who died in 2007 at the age of one hundred. More than seventeen hundred plant varieties thrive there now, and the arboretum is open to visitors and researchers.

SAGAPONACK SCULPTURE FIELD on Long Island, New York, holds the collection of Louis Meisel, a Soho art dealer and longtime tree collector with a focus on beeches. He likes the trees because they are able to resist the strong hurricane winds, aren't much bothered by pests or disease, and leaf out in a beautiful range of colors: purple, copper, pink, and green, turning to gold and orange in the fall. The public is welcome to visit his collection of trees and sculpture. Over the years he has also collected antique lawn sprinklers, vintage ice-cream scoops, and old enamel signs. "Trees are the only thing I collect that I don't also sell," he said. "Sometimes people ask me what kind of trees they should plant, but they just want one impressive specimen to show off in front of their house. They're not collectors."

STARHILL FOREST ARBORETUM in Illinois, founded by Guy and Edie Sternberg, is known worldwide for its extraordinary collection of oaks. Starhill Forest also serves as the official arboretum of Illinois College, providing research and education opportunities.

CERTIFIED TREE COLLECTIONS

ArbNet is an international accreditation program for arboreta. Both public gardens and private collections are eligible. Certification involves documenting the collection, adhering to certain practices, and offering some level of public access. Around the world, many countries have their own "national collection" certification programs that tree collectors can also participate in. These programs are strictly voluntary, but it's a nice way to get public recognition for a lifetime of tree planting.

"I STARTED LOOKING UP THE NAMES OF OTHER FRUITS I'D NEVER HEARD OF BEFORE. AND I NEVER STOPPED."

THE RARE FRUIT COLLECTOR

HELTON JOSUÉ TEODORO MUNIZ

Campina do Monte Alegre, São Paulo, Brazil

WHEN HELTON JOSUÉ TEODORO MUNIZ WAS FOURTEEN YEARS old, his family moved in with his grandparents to help them run their farm in a rural area of São Paulo. It was there that he learned about saputá (*Cheiloclinium serratum*), a rare indigenous fruit found only in that part of Brazil. The yellowish fruit, about the size of a plum, is delicate and sweet, and its seeds can be roasted and eaten.

"I was curious and went to look up the name," he said. "My grandmother had a dictionary of several volumes. I started looking up the names of other fruits I'd never heard of before. And I never stopped."

After his grandparents died, Muniz stayed on the farm. He started exchanging seeds with other collectors through a magazine called *Globo Rural,* where readers could list the seeds they were looking for. "After the internet came, it got easier," he said. He found other collectors through online groups and created a website to list the seeds he had available. He set out by foot as well, walking miles through untamed wilderness to find fruit he'd never seen before.

A disability caused by a lack of oxygen at birth has limited his mobility somewhat: when he's in pain, he relies on a garden tractor to get around, and he has help from his wife, his father, and a few employees. "The pain gets in the way," he said, "but working with fruits gives me good health."

Muniz works every day in his fields and greenhouses, and also in his office, where he's surrounded by reference books and stacks of plastic

Saputá *(Cheiloclinium serratum)*

cartons of seeds he's collected. He's self-taught, but university researchers come to him to learn more about the wild fruit he cultivates.

Now, in his forties, he's filled the farm with an astonishing array of native Brazilian fruits, including the many species of pitanga (*Eugenia sp.*), a tiny red fruit in the myrtle family that he can only describe as "unforgettable," and the very rare vine watermelon, *Calycophysum weberbaueri,* with fiery orange fruit that tastes like its namesake.

He cultivates thirteen hundred species, plus another three hundred exotic fruits from around the world. "Brazil has four thousand species of fruit," he said, "so I have twenty-seven hundred to go, and I'm out of room to plant more. And of course, each species has its variants. This is an infinite world. I would need a lot more land."

It's a remarkable collection that draws fellow collectors from around the world. But he most wants to educate the locals. They're always astonished to see vaguely familiar, nondescript shrubs and trees cultivated on his property in a naturalistic setting, not too different from how they might grow in the wild. "People say, 'But this is bush.' They just don't know. I see mothers and fathers telling their children, 'Don't pick this fruit, it's poisonous.' In fact, they simply don't know what the fruit is. I understand that they want to protect their children, but it's wrong information. It's a lack of knowledge."

Guabiroba do campo
(Campomanesia adamantium)

Most of the fruits that interest him grow wild in the tropical savanna known as the Cerrado in central Brazil, and in what he describes as "the fields of unknown Brazil." And no one pays much attention to them.

"Brazilians tend not to value what they have in Brazil," he said. It wasn't until rare fruit collectors from around the world started to take notice of the species he was growing that Brazilians started to pay attention too. "Once they value it abroad, Brazilians don't want to be left behind," he said.

Muniz has published a two-volume reference book on the native fruits of Brazil and posted in-depth descriptions and photographs of more than five hundred species on his website. He also gives talks at rare fruit symposiums, sells both seeds and seedlings, and makes videos about the cultivation and uses of the fruit.

The highlights of his collection are fruits that few people outside of Brazil have ever seen. Guabiroba do campo (*Campomanesia adamantium*) produces grape-sized, lime-green fruit with a soft, juicy pulp that makes delicious jelly and ice cream. Cereja do cerrado (*Eugenia calycina*) is a small, bright-red fruit that tastes like a cherry and is getting hard to find in the wild.

Many of these fruits can't be easily transported to market, but they can be dried or made into jelly or juice and sold. That's his next project. "In this part of Brazil, they grow beans, corn, and soybeans," he said. "I want people to understand that fruit can be grown more profitably than soybeans. The ecology is important, but we have to think about the profits too. The economy has created the environmental issues we have today."

Cereja do cerrado
(*Eugenia calycina*)

"FOUR IS BETTER THAN ONE. BUT FORTY WOULD BE EVEN BETTER."

THE INDEPENDENT RESEARCHER
DEAN NICOLLE
Adelaide, Australia

DEAN NICOLLE GREW UP AROUND HIS PARENTS' CYMBIDIUM OR-
chid nursery, but the tropical flowers never interested him. What he noticed were the eucalyptus trees around the edge of the property. The
trees are ubiquitous in Australia, but they're otherworldly and fascinat-
ing to anyone who takes the time to look closely. With their striped,
papery bark in a range of hues from gray to pink to blue to shocking
orange, oddly shaped seed pods that resemble miniature flying saucers,
flowers that bloom like tiny fireworks displays, and the sharp mentho-
lated fragrance, there's a lot in every tree to fascinate an inquisitive kid.

His parents encouraged his interests: instead of buying him comic
books they gave him a field guide to eucalypts when he was only eight
years old. "I read every page until I knew them by heart," he said. "I'd go
around the local suburbs and try to identify as many as I could. It wasn't
long before I was trying to get my parents to take me to the nursery. Any
eucalypt with a different name on it, I'd ask if they'd buy it for me. I'm
really lucky that I had parents who took an interest in my interests."

Buying a book and a few plants is one thing, but then Nicolle's parents spotted a good deal on an eighty-acre parcel of land near their
home in Adelaide, Australia. "They could see that if I was going to continue to plant trees on their property, they'd never be able to sell up and
retire," he said. "So they bought this land when I was about sixteen. I
had a dream that I wanted to grow every eucalypt species that there
was."

Corymbia ficifolia

To invest in a teenager's dream takes quite a leap of faith, but in this case it paid off. Nicolle attended a specialized high school with a focus on horticulture and then studied botany in college. Many students struggle to choose a major, much less a career. He had his life's work waiting for him when he graduated.

"I knew what I wanted to do," he said, "but I never knew how I was going to make a living at it. It was a passion."

He started planting his arboretum in 1992, when he was still in college. Most tree collectors begin with little more than an amateur's enthusiasm, but Nicolle had the benefit of his formal education. He understood what it would take to establish a useful research collection. He planted his trees in blocks by year, organized them into rows, assigned a number to each one, and cataloged them in a database.

Eucalyptus conferruminata

He named the place Currency Creek Arboretum. It's the world's largest collection of eucalyptus species, with almost ten thousand trees planted. Nicolle has come close to meeting his goal of growing every eucalypt in the world: at the moment, there are about a thousand named species and subspecies, although that number can change as new ones are discovered or reclassified. "I'll probably never get to one hundred percent," he said, "but I've been close. I have grown or tried to grow nine hundred fifty of those."

Eucalyptus brandiana

He made a decision early on to plant only trees grown from seed he'd collected himself. From each mother tree he selects in the wild, he takes

Eucalyptus sinuosa

a voucher specimen, which is a pressed herbarium record that can definitively identify the tree from which the seeds were harvested. Then he germinates the seeds and plants four siblings in a row together, so that they can be compared for genetic variations even as they are grown in nearly identical conditions. He'd plant more than four from each mother tree if he had the space and the time. "Four is better than one," he said. "But forty would be even better."

Almost all eucalypts are native to Australia, so he's done most of his seed collecting in his home country. "There are a few species I haven't been able to get hold of because they're growing in very remote areas, or because they only seed at a particular time. One of the ghost gums, *Corymbia aparrerinja,* sheds its seeds as soon as the fruit ripens, all in the course of a week. This is in the tropics, during what we call the wet season. You can't really get to them in time. Those are the challenging ones, but they're the ones that are fun to try to collect."

Eucalyptus forrestiana

Nicolle has built a remarkable career for himself around his lifelong

Eucalyptus macrocarpa

passion. He teaches, writes books, gives lectures, and offers his services as a consulting arborist. He's published and given names to 102 previously unknown species and subspecies, and he continues to look for more.

But his long-term research makes Currency

Eucalyptus caesia

Creek more than just a collection. Right now he's researching the effects of wildfires. "They're increasing in frequency and intensity in Australia. We need to know how different species respond to fire and what it takes for them to regenerate. So we can burn up a block of trees and study which ones come back. All the local fire services come out and use it as a training exercise. With a private arboretum, you can do that sort of destructive research."

He also hopes to encourage Australians to appreciate the diversity of native eucalypts. "They're not all Tasmanian blue gums. There's so much variety, in terms of the flowers, and the colors in the bark, and really small eucalypts that host so much wildlife. I'd like people to look at them in that light."

Eucalyptus stenostoma

Corymbia ficifolia (flower)

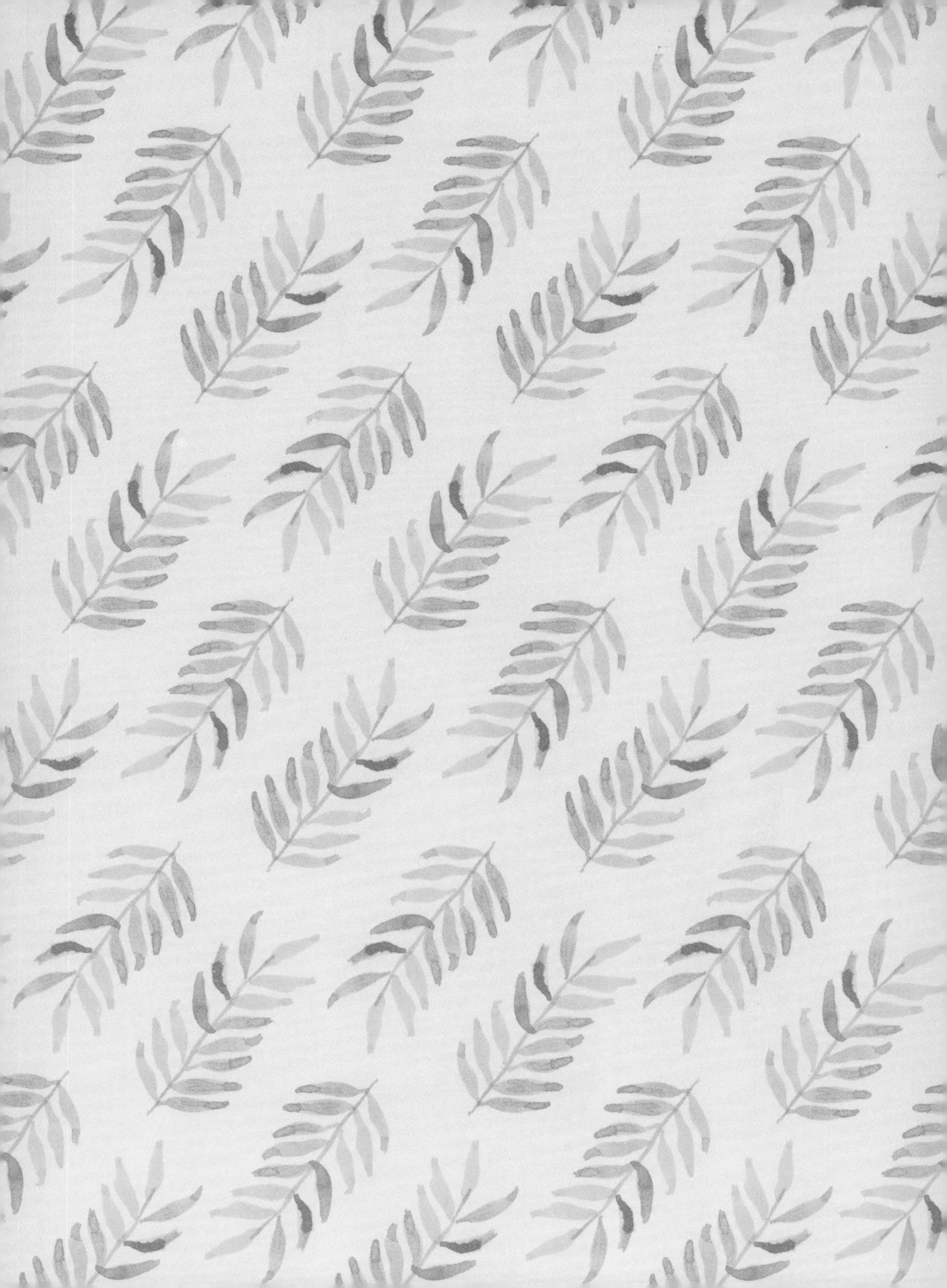

"FRIENDS ARE ALWAYS TELLING ME I HAVE TOO MANY TREES. THE GREAT TREE EXPERT MICHAEL DIRR HEARD THAT FROM HIS FRIENDS TOO, AND HE SAID, 'OBVIOUSLY, FRIENDS LIKE THESE ARE EXPENDABLE.'"

THE INTERCONTINENTAL COLLECTOR
FRANCISCO DE LA MOTA
Madrid, Spain, and Houston, Texas

WHEN FRANCISCO DE LA MOTA WAS ABOUT TWELVE YEARS OLD, he picked up one of his parents' books about trees. "It was Hugh Johnson's book. We had it in Spanish, but in English it was called *The International Book of Trees*," he said. "It was full of these fantastic photos, and the way he talked about the trees was very appealing, because he wrote about how they lived in their natural environment. He wrote about ginkgoes and how they had survived virtually unchanged for two hundred million years. And I said to my parents, 'I want these trees in my yard.'"

His parents owned a vacation home north of Madrid. They indulged his whims and bought him a ginkgo. When he grew up, he started to live year-round at a property next door, which allowed him to expand his collection. "The tree selection in Spain was limited," he said. "The internet opened up the possibility of exploring nursery catalogs from the UK, from the Netherlands, France, anywhere I could get them." In 2002 he started a landscaping business, which gave him access to wholesale nurseries and made it even easier to order trees for himself.

He was particularly interested in maples for their fall color, and birch trees for their unusual bark. He joined the Maple Society and started trading rare and interesting trees with fellow collectors. A highlight of his collection is the endangered *Acer pentaphyllum,* a small, shrubby tree with leaflets arrayed around the stem in groups of five, like fingers on a hand. Only about five hundred of these trees still live in the wild, in

China, but collectors around the world have been propagating them for decades. "My tree is nearly twenty feet tall, and it's producing a lot of seed," he said, "so I can exchange seed with maple lovers now."

He created a spreadsheet to keep track of his collection, recording where he bought each plant, as well as blooming times and other key dates. "I started treating my collection like a botanical garden," he said. But it's not a botanical garden—it's an ordinary garden, and space is always a problem. "I only have half an acre. If you're collecting succulents, that could be a decent space. But with trees, it's not enough. I overplanted. Everything is too close together and too close to the house."

Perhaps it was inevitable, then, that he would take his collecting impulse to an actual botanical garden. After working as a tree consultant for the city of Madrid and earning a PhD in horticulture from Virginia Tech, he has joined the Houston Botanic Garden as its director of horticulture. Here he can grow tropical maples that just wouldn't thrive in Madrid, as well as endangered oaks, magnolias, and tropical conifers. "And our palm collection! There's a passion here in Houston for anything tropical. But I think they would like to see some fall color too."

While he's introducing unusual trees to Houston, and trying, with the help of family, to keep an eye on his collection back in Madrid, there's yet a third plot of land where he's managed to introduce a few trees. "I met my wife when we were both doing our internships at the Holden Arboretum outside of Cleveland. She grew up on a farm in Northeast Ohio, and her family still has four hundred acres. So, of course, I planted a few trees there. I put in *Franklinia*, dogwoods, and some redbuds. They have a pond, so I was able to plant several bald cypresses, because those love the water. I've always been fascinated with them. But I try not to plant too many trees there because I just don't get there often enough."

His only regret is that he doesn't have even more space to expand his collection. "Friends are always telling me I have too many trees. The great tree expert Michael Dirr heard that from his friends too, and he said, 'Obviously, friends like these are expendable.'"

"IT WILL SURVIVE. THAT'S
A GREAT REASON TO
LOVE A TREE."

THE FIFTH GENERATION
COR VAN GELDEREN
Boskoop, the Netherlands

COR VAN GELDEREN DIDN'T INTEND TO JOIN THE FAMILY BUSI-
ness. "I was planning to become the best Dutch novelist, but I did not
have the talent to become the best one, only the second best," he said,
laughing. "I decided when I was twenty-five years old that growing trees
and shrubs was an interesting thing to do as well."

He's the fifth generation of Dutch tree growers in his family. "Usually
the first generation starts it, the second generation builds it up, and the
third generation ruins it. But I am the fifth! My children are not interested in horticulture whatsoever. I have six children. You can't say I
didn't try. But I don't know if there will be a sixth generation."

Perhaps the impulse will come to them in adulthood, as it did for van
Gelderen. On a trip to an arboretum in Rotterdam, he caught the scent
of a familiar rhododendron. "It brought me back to my youth, when I
used to go with my father collecting cuttings and putting them in bags.
It just came back to me, what a good memory it was. And I decided to
stop trying to get away from my karma. That was when I joined the family business."

The urge to collect trees also came to him by accident, after visiting a
library to browse a collection of nursery catalogs from the nineteenth
century. "These nurseries had enormous collections of boxwoods I'd
never heard of. I got intrigued. Where had all those boxwoods gone?
They must be somewhere. So I started looking for them, and eventually
I had twenty different boxwoods. But I got bored with them quickly."

The next opportunity came along when his father published a book on maples. "He worked on it for fifteen years. It was quite a significant book. But when it was published, the only comment I had was that there weren't so many pictures in it. I could see he was really disappointed in my reaction. This was his life's work! So I asked if we could write a book together, one that had pictures in it. That's when I started collecting maples. Then it was hydrangeas, and then I started a camellia collection. And I got bored of those too."

Tree collections take up space, even in pots. The family nursery sits on only nine acres. "It's quite a good size by Dutch standards, but it's small compared to an American nursery," he said. "I'm running out of room. Everything's on top of everything else. Now I'm afraid I'm about to start collecting birches. I hope I won't, because there are so many birches. But I probably will. Once I get interested in a certain tree, I want to know everything about it, and I need to see the trees. I can't do it just by reading books. That's my problem, really. And there's no cure for it."

Over the years he's noticed that some collectors go for beauty, some for rarity, and some just like a tree with a good story. The Nippon maple, *Acer nipponicum,* checks a couple of those boxes. "You wouldn't collect it for its beauty because it's not beautiful at all. It looks a bit like a sycamore, but uglier. It doesn't have good autumn colors. The best thing you can say about it is it's really, really rare. There was only one in Holland, in an arboretum in the Hague, and they chopped it down because they thought it was a really common maple. Only two years before that, we'd collected seeds from it. It was the first time it had ever produced seeds, so we were lucky to get any at all. We grew some small trees from that, so this maple is still alive in Holland. But nobody wants one, because it's not a good-looking tree. Rarity is not enough."

But there are other trees he loves as a collector because they are so common. "They soften my heart," he said. "There's a maple you see everywhere in Europe, *Acer tataricum*. In Germany, gas stations along the motorway are called Raststätte, or rest place. So we rebaptized *Acer tataricum* to *Acer raststatteanum*, as this is the place that you will find this maple most often. It can withstand everything: poor soil, pollution, and gas fumes. It will survive. That's a great reason to love a tree."

"I FOUND OUT THAT I COULDN'T TELL THE DIFFERENCE BETWEEN VARIOUS TYPES OF WOOD, AND THAT BOTHERED ME.... THAT'S WHERE IT ALL BEGAN."

THE WOOD COLLECTOR

DENNIS WILSON

Alpena, Michigan

DENNIS WILSON'S TREE COLLECTION DOESN'T SPAN SEVERAL acres, nor does it require regular pruning or irrigation. His trees are stored in boxes and cabinets in the wood room attached to his workshop.

"I started collecting wood samples as a hobbyist woodworker," he said. "I was refinishing furniture, and if I had to make a drawer or another piece, I wanted the wood to match. But I found out that I couldn't tell the difference between various types of wood, and that bothered me. So whenever I made something, I would cut myself a small sample of that wood and put it on a shelf so I'd have a reference piece. That's where it all began."

Then he discovered the International Wood Collectors Society (IWCS) and became a member. "They'd send out pages and pages of listings for wood samples from this guy named Art Green. He had twenty different types of maples alone! I didn't know what any of those looked like. Well, at the time these samples were twenty-five cents, maybe a dollar, so I ordered a box of them." He met even more kindred spirits through regional and national meetings and wood swaps.

Wilson, who is now president of IWCS, recalls the early days of the organization. "When it was started back in 1947, it was mostly a group of dendrologists who were really technical people. They were very interested in the structure of wood. Back then, you could write to a forestry department in many places around the world, and they'd send you samples of wood from their local trees. But those samples all came in different sizes. You never knew what you were going to get." In response, the IWCS established a standard size of three inches by six inches by half an inch, which is most commonly used by collectors today. When every sample is a little larger than a smartphone, it's easy to store, organize, catalog, and display a collection.

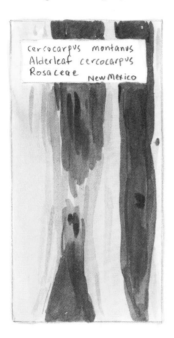 

Today Wilson has about 6,900 wood samples, mostly of the standard size, labeled with the tree's common name, botanical name, and location where it grew. His is one of the largest collections in the world outside of a public institution.

The largest collection of wood samples in the United States is not far from Wilson's home in Michigan, at the USDA's Forest Products Laboratory in Madison, Wisconsin. It houses 105,000 wood samples, which includes its own collection and those it acquired from other institutions, such as Yale University and Chicago's Field Museum. "They inherit these collections when other institutions close their research departments," Wilson said. "Of course, they have duplicates. They don't want just one sample of a tree. They want a number of different samples from the same species, grown in different places. They're looking at the anatomy of the wood and comparing the structure depending on where it's grown."

Wilson's career began with a different sort of woodworking: he got his start in the automotive industry, creating wood patterns for engine parts. "I would make the shape of engine parts out of wood. They'd take that and make a metal pattern from it, and then it would go to the foundry to be made into a mold. They started phasing out those wood patterns twenty-five years ago, but some manufacturers still do it that way. It can be quicker and less expensive than designing it on the computer if you know what you're doing."

Today he makes wooden toys for his grandchildren, along with tables, beds, and nightstands. He's working on a more elaborate set of custom-built cabinetry to house his wood collection, which has sprawled into dozens of cardboard boxes. He's inherited Art Green's collection of boards, which he cuts into samples and sells to raise money for the IWCS. It makes for quite a lot of wood in need of organization. "I have an understanding wife," he said, "but there are limits."

REMARKABLE WOOD COLLECTIONS

Xylarium

XYLARIUM • A collection of wood specimens (from the Greek *xylo*, meaning wood, and Latin *arium*, meaning a place associated with a particular thing—think herbarium, aquarium, planetarium).

INDEX XYLARIORUM • A list of wood collections around the world, compiled by American botanist William Louis Stern in 1967, now published as an updated online database through the Global Timber Tracking Network.

WORLD'S LARGEST WOOD COLLECTIONS

185,647 specimens   *Xylarium Bogoriense*
                    in Bogor, West Java, Indonesia
125,000 specimens   *Naturalis Biodiversity Center,*
                    in Leiden, the Netherlands
105,000 specimens   *USDA Forest Products Laboratory,*
                    in Madison, Wisconsin
84,600 specimens   *Federal University of Pernambuco*
                    in Recife, Pernambuco, Brazil
83,000 specimens   *Royal Museum for Central Africa*
                    in Tervuren, Flanders, Belgium

XYLOTHEQUE • A book-shaped volume of wood samples, leaves, or twigs, traditionally made of wood and displayed in a cabinet of curiosity. The most famous examples are:

German forestry professor HERMANN VON NÖRDLINGER published a remarkable series of books called *Querschnitte von hundert Holzarten* (or *Cross Sections of One Hundred Wood Species*) between 1852 and 1888. Each volume consisted of a wooden box containing one hundred thin slices of wood samples mounted on sheets of paper and accompanied by a pamphlet describing the samples. He produced eleven volumes in the series, spanning a remarkable eleven hundred species of conifers, palms, broadleaf trees, cycads, and tree ferns. Five hundred copies were produced, and the complete set can still be found at a few libraries around the world.

Between 1894 and 1928, American botanist ROMEYN BECK HOUGH published the fourteen-volume *The American Woods: Exhibited by Actual Specimens and with Copious Explanatory Text*. Each page presented three real slices of wood from a North American tree, cut from three different angles, accompanied by descriptive text. He collected the wood samples himself and documented 354 species but died before he could complete the work for what would have been the fifteenth volume.

Japanese scientist **MOGAMI TOKUNAI** created a collection of forty-five wood samples in the early nineteenth century, all neatly packed into a custom-made crate, with delicate paintings of each tree's leaves on one side, accompanied by descriptive text. He gave this remarkable collection to Dutch scientist Philipp Franz von Siebold in 1826. Today it's held at the Naturalis Biodiversity Center in Leiden, the Netherlands.

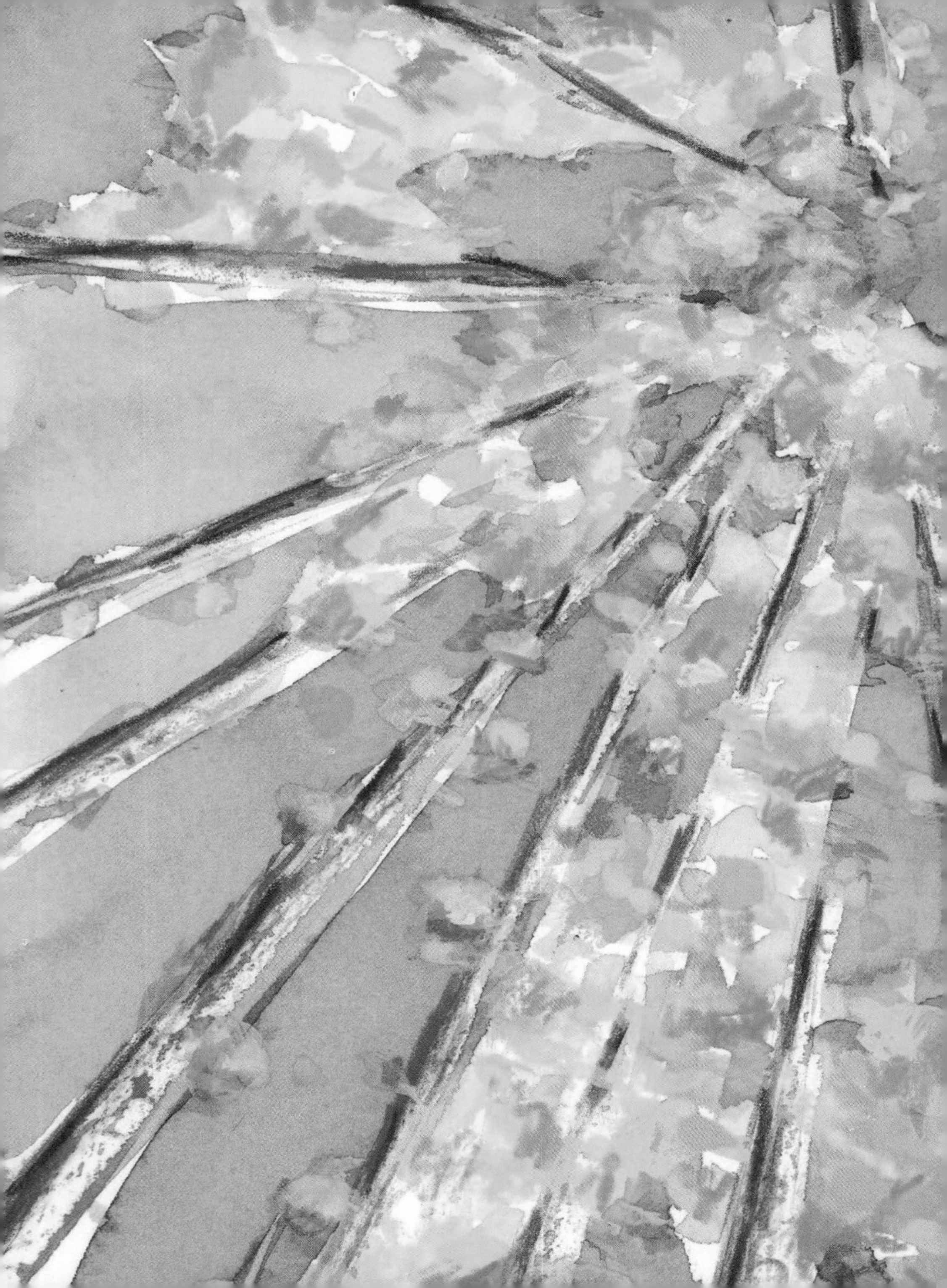

EDUCATORS

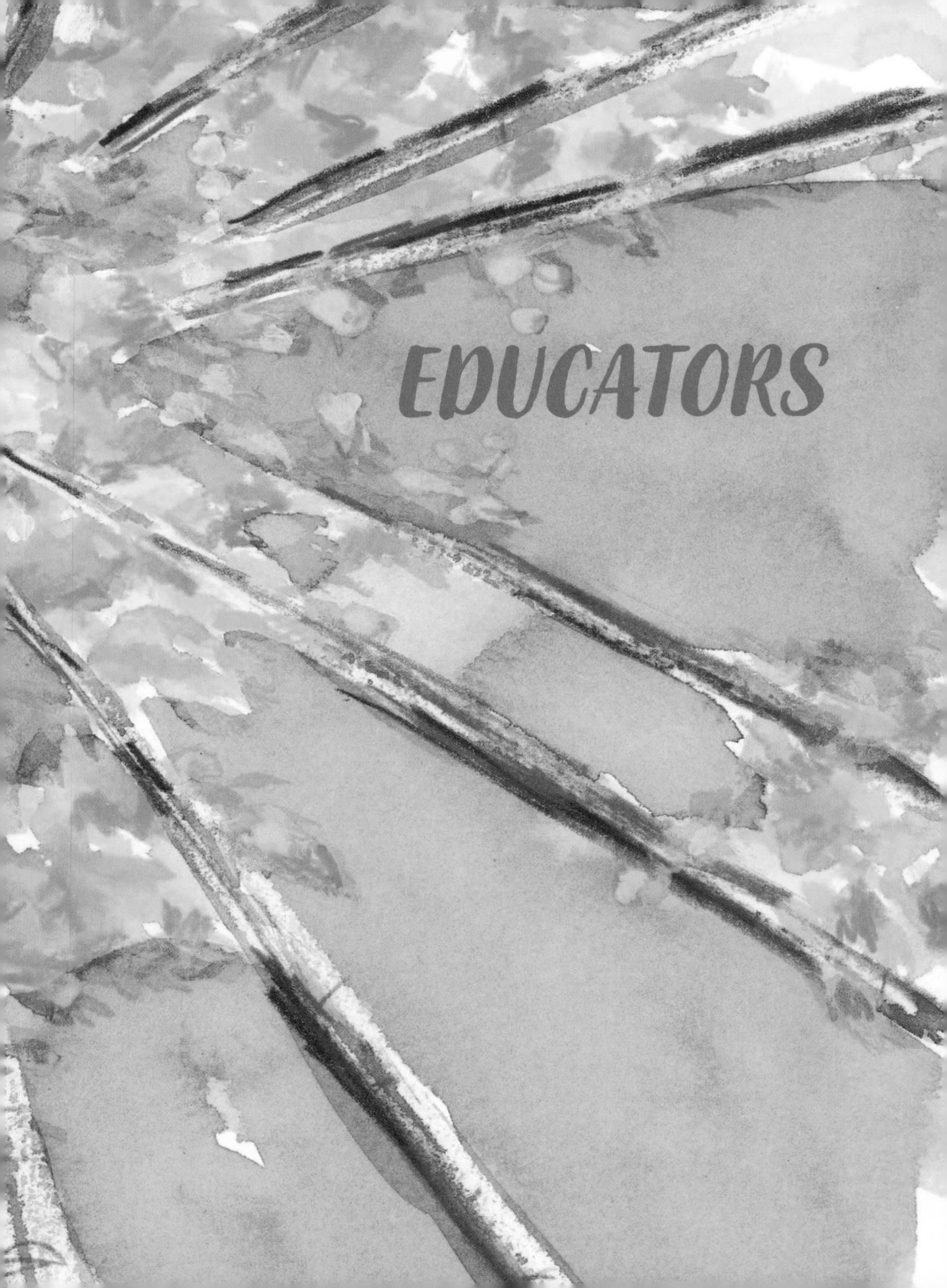

"*TREES OF STANFORD* WAS PURELY AN EXPRESSION OF DELIGHT."

THE CATALOGER

SAIRUS PATEL

Palo Alto, California

SAIRUS PATEL, A RETIRED FONT STRATEGIST AT ADOBE, BECAME interested in urban forestry about fifteen years ago. He started attending tree walks in his hometown of Palo Alto, California. "I grew up in India, where I knew, at least, the trees on the streets there," he said. "So I wanted to learn about the local trees. I started taking classes and looking at the trees on Stanford's campus, where I'm an alumnus, and then a friend gave me this book."

The book was *Trees of Stanford and Environs,* written by Ron Bracewell, an internationally renowned professor of electrical engineering and astronomy who taught and lived on campus throughout the latter half of the twentieth century. When he wasn't developing the technology that powers CAT scans or creating a radio telescope to support NASA's moon landings, Bracewell was walking around campus noticing the extraordinary diversity of trees around him. He began researching both the botany and the history of more than forty thousand trees growing on campus. "It was almost as though there was no practical focus to his project," Patel said. "*Trees of Stanford* was purely an expression of delight."

He found plenty to delight him. The land on which Stanford sits was purchased by industrialist and former California governor Leland Stanford in 1876. He hired Frederick Law Olmsted to design an arboretum, landscape plan, and overall design for what would become Stanford University. While the campus has changed a great deal since it opened in 1891, its tree collection is a remarkable testament to the varied interests of university botanists, arborists, and groundskeepers over the decades. Unusual conifers, exotic eucalyptus, native oaks, and tropical palms can all be found among the four hundred species on campus.

In 1972, Bracewell published a spiral-bound, photocopied directory of campus trees, which he illustrated by gathering leaf specimens, placing them on the copier, and reproducing them directly. The size of the leaves dictated the size of the book, which, even in its fourth printing as a paperback, is still only slightly smaller than a sheet of copier paper. "His writing was opinionated, quirky, and wry," Patel said, "and the book was quite charming, not a dry botanical treatise."

Patel was fascinated by the book and by the ongoing effort to research and catalog Stanford's tree collection. John Rawlings, a librarian on campus, had inherited the project from Bracewell and developed a

Trees of Stanford website. In true librarian fashion, Rawlings added citations, reference notes, and an extensive bibliography. The site serves as an archive as well as a directory: even long-dead campus trees are documented.

The meticulous nature of this work, combined with the clear sense of affection for the trees, appealed to Patel. "In 2013 I approached John about updating the website," Patel said, "and at some point, he said, 'Sairus, it's yours.'"

As the third caretaker of the Trees of Stanford project, Patel is at work on a new edition of the book. He helps to teach interdisciplinary classes on campus that touch on themes of botany, conservation, and the history of campus trees, and he leads tree walks.

But he hasn't forgotten his career in typefaces. Just as John Rawlings's work as a librarian connected to his interest in cataloging trees, Patel feels that his obsessive interest in the details of fonts translates to trees. "I used to go into bookstores and open books upside down to see if I could identify typefaces in a couple of seconds. It's the same with trees. When you're identifying the different species of eucalypts, you have to look at the very small details of buds and groupings. There is definitely, for me, a bridge between fonts and tree identification."

HOW TO COLLECT WITHOUT OWNING A SINGLE TREE

You don't have to plant a tree to be a tree collector. The act of drawing a boundary around a group of trees, and cataloging and naming them, is a way of building a collection. This is both the charm and value of the Trees of Stanford project: it defines the trees on campus as a collection and provides a directory that explains the significance of each tree. Matt Ritter, a professor of biology at Cal Poly San Luis Obispo and avowed tree lover, said, "As soon as you put a label on a tree, that changes everything. It's a lot harder to cut it down. I make sure every tree on campus has a label."

A tree-cataloging project doesn't require a big organization, or even a committee. Here are just a few projects started by passionate individuals:

LONDON • Paul Wood remembers collecting trees as a child, which is to say that he dug up tree seedlings from a wooded lot and planted them in yogurt cups. His tree-cataloging efforts started as a blog about London's street trees and evolved into a series of books on the trees of London. He also leads tree walks and helps to organize an urban tree festival, and now he's involved in a digital tree mapping project called Tree Talk. His favorite London trees include the enormous churchyard yew in Downe, where Charles Darwin lived, and the ancient Cheapside Plane, whose protected status limits the height of any nearby building that would interfere with its limbs.

NEW YORK CITY • Jill Hubley started taking walks in Prospect Park, Brooklyn, to escape the confines of a small apartment. She'd always loved field guides and started looking around for a guide to the trees in the park. She couldn't find one, but she did discover that the city of New York had been taking a street tree census every decade since 1995. Her  skills as a web developer got her started on a digital map of the city's trees based on the census data, but she had to learn data visualization along the way. Her map, "NYC Street Trees by Species," is a colorful depiction of New York's tree canopy, color coded by species. "I get emails from people who are like, 'That tree is not there anymore,'" she said. "I can't go through and update it tree by tree. But people are very invested in the trees on their streets."

 **LANCASTER COUNTY, PENNSYLVANIA** • Len Eiserer, profiled on page 205 for his own tree collection, started a website called Tree Treasures of Lancaster County, in which he catalogs "those trees in Lancaster County that are in some sense 'special'" as a way of paying tribute to "all trees in our county, our state, our world." He's cataloged more than three hundred tree treasures and takes nominations from the public.

MEXICO CITY • Francisco Arjona went to college to study engineering, but it didn't feel right. A job at the university's arboriculture lab ignited his passion for trees. "When I started working with trees, it was like, this is what I want to do. I have ideas, I have energy, I have emotions. I knew I wanted to work with trees my whole life."

The inspiration for his Trees of Mexico City (Árboles de la CDMX) project struck him in the middle of the night. "I just wanted to see some trees. But it was totally dark out. I started looking online, and nobody was posting about the trees here." Now he shares his favorite trees on Instagram, TikTok, and Twitter and looks forward to leading tree walks and building a tree map of Mexico City as he finishes his master's degree in forestry science. "There's no conservationist culture about trees here," he said. "But even if you just know the name of a tree, it starts to mean something to you."

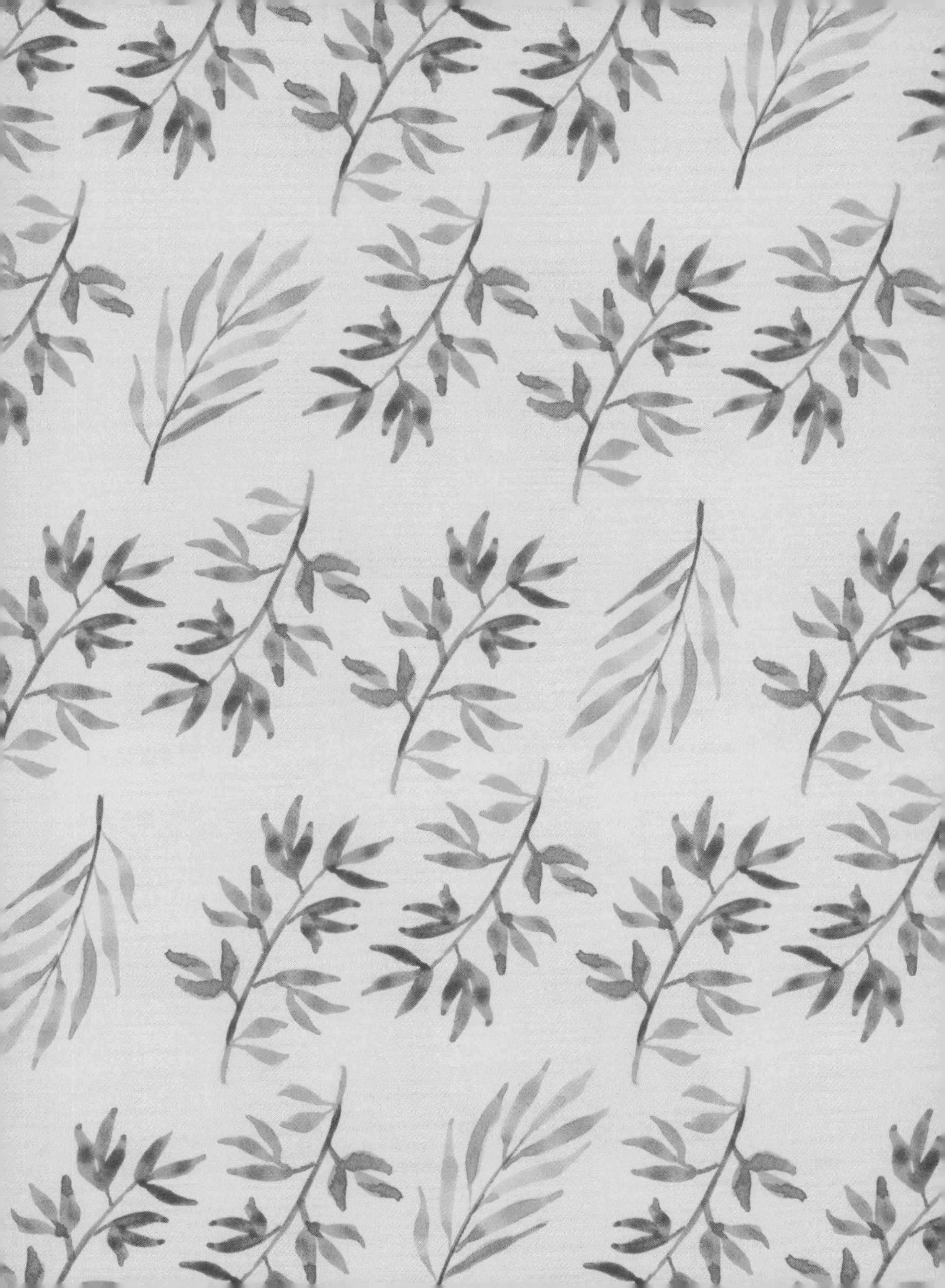

"THESE TREES ARE KIND OF A MENTAL BREAK FOR ME.
JUST LOOKING AT ALL THE GREENERY, IT'S RELAXING."

THE TROPICAL YOUTUBER

KAO SAELEE

Visalia, California

IT'S EASY TO FIND KAO SAELEE'S HOUSE. ON A STREET OF MANI-cured suburban lawns in California's Central Valley, his home is almost entirely obscured by a tropical jungle. Mangos and bananas are on offer to passersby. Birds rarely seen elsewhere in the neighborhood are coming in for a landing. Kao himself might be outside on the sidewalk, in front of a video camera, telling his YouTube followers how it is that he's managed to grow 168 tropical trees on a standard suburban lot.

What's even more remarkable is that he's only been at it for about five years. "It was actually YouTube that got me started," he said. "One evening, I saw a video from a gentleman in Minnesota who was trying to grow an orange tree in his cold climate. That's a lot of effort, just to keep one little tree alive! But that made me dig deeper, to see what it would take to grow some tropicals here."

Saelee's family immigrated from Thailand when he was nine. "I remember quite a bit of Thailand. One of the reasons I got into growing tropicals is that many of these trees grow there. It goes back to my childhood, and bringing some of that here."

The first tree he planted was an atemoya tree (*Annona squamosa* × *A. cherimola*), which is a hybrid bred in Florida in 1908 that has gone on to be wildly popular in Thailand for its custardy fruit with notes of banana, pineapple, and vanilla.

"I planted that one in the ground, not in a pot," he said, "and I was really surprised over the next couple of winters at how well it adapted.

From there, I went into the more advanced trees, like mangos, that would absolutely need sheltering when the temperatures got down to freezing."

One tree led to another, and soon he was making trips to a tropical tree nursery in Los Angeles for harder-to-find varieties. Both mangosteen, a segmented fruit about the size of a tangerine with a thick purple peel, and rambutan, a small white fruit similar to a lychee, covered in a spiny red peel, made it into his collection. He also grew the ice cream banana tree, often called 'Blue Java' for the distinct light-blue color of the banana peel before it ripens.

"Given the size of my yard," he said, "it made sense to concentrate more on edibles that would give me something." Almost all of his trees live in containers so that he can move them to shelter if winter temperatures get too low. Growing in containers also keeps the trees from getting too large and crowding each other out. "If I were to put these trees in the ground, they would get so big that they'd cover the other trees.

This just allows me to have so many more. I would love to have them all in the ground if I had the space."

His wife encouraged him to start posting videos on YouTube about growing tropical fruit in California's Central Valley. It's not the easiest place to grow a tree that is accustomed to Thailand's warm, damp climate: summers are relentlessly hot and dry, and temperatures dip below freezing in winter. He posts a video once a week or so, cataloging his failures and successes, and a chorus of viewers weighs in to offer advice or ask questions.

Thanks to his videos, he's found a community of like-minded collectors, both around the world and nearby. "I've found quite a few fruit growers in my area," he said, "and we've exchanged information and seeds and cuttings. Sometimes I'll give tours to local viewers, and I give them a starter kit with seedlings that I grew. These exchanges with locals have been tremendously helpful."

His trees produce more fruit than his family can eat, but they also offer him a respite. "I work in IT, which is as disconnected from trees as you can get. There's a lot of stress, working in the tech industry. These trees are kind of a mental break for me. Just looking at all the greenery, it's relaxing."

In the middle of summer, with the garden at its most abundant and the leafy canopy overhead creating its own cool, damp microclimate, it can seem as though his tropical fruit forest has been there forever, but he knows how far it has to go. "Everything you see here is just the beginning," he said. "In human age, this garden is just learning to crawl."

"A SINGLE CONE IS A WAY TO LOOK AT THE WHOLE NATURAL HISTORY OF PINE TREES."

THE PINE CONE COLLECTOR

RENEE GALEANO-POPP

Sapello, New Mexico

In 2009, Renee Galeano-Popp retired from her job as a botanist and forest ecologist. She could see right away that she'd need a project in retirement.

"My husband was still working as a forester at the time," she said. "I don't remember exactly how it came up, but somehow one of us said, 'You know, there's only about a hundred and fifteen species of pine around the world. What would happen if I tried to collect the cones?'"

Pine cones are remarkable bits of botanical engineering: they stay tightly sealed to protect seeds from winter weather and attacks from predators, then open and release the seeds when conditions are right for the next generation to germinate. Some species depend on wildfires to release the seeds, while others produce long, grasslike needles around the growing tips to guard against fire. Most of all, they're durable. Throw a few in a box, and you've started a collection.

This turned out to be a wonderful excuse for a post-retirement road trip. "I set out in my truck and spent a summer in the West—California, Utah, Colorado, Nevada—and I collected as much as I could."

That yielded cones from more than a dozen western species. She then reached out to the major arboreta around the country and asked if they would contribute to her collection. "I didn't want cultivars, only straight species," she said. "They obliged me with about another dozen species."

From there, she created a Facebook page called Project Pine Cone.

People sent her cones from around the world. A collector in Germany was eager to trade, as was a forestry student in El Salvador.

"My collection was growing and growing. Then a colleague of my husband's at the Forest Service died and left his cone collection. Nobody knew what to do with it, so they all said, 'Well, let's give it to Renee.' These were Smithsonian-quality specimens."

One of the prized cones from this collection was *Pinus maximartinezii*, also called the maxipiñon, which grows only in a remote mountain range in Mexico. It had not been described in botanical literature until the 1960s, when a Mexican botanist noticed unusually large edible seeds for sale in a market and located the tree that produced them. Today it enjoys protected status in Mexico and is sometimes sold as a rarity from specialty tree dealers. The cone alone is a dramatic specimen, reaching nine inches in length.

With such impressive additions to her collection, she created a display that could be taken around to classrooms. "Word got out, and people would contact me and ask me to come share the collection. I started doing workshops for native plant societies and volunteering for the Colorado State University herbarium. I donated a display of cones from my collection, all arranged in taxonomic order. They made a floor-to-ceiling glass display of twenty-five species of cones."

That project occupied the first twelve years of her retirement. In that time, she amassed cones from seventy-seven species of pine, along with some from fir and larch. Within an individual species, there are variations between the cones depending on the region or environment where the tree grows. A cone might be longer or shorter. The scales might be curved or straight. When she could, she would gather examples of these natural differences.

She's no longer actively collecting, having gathered as many as she can without trekking through remote areas of Mexico and Asia. "I have a friend who's living in Cambodia, and I asked him if he could get me some cones. He said, 'There's so much illegal logging that they put armed guards around their forestry operations. If it's not them, it's monkeys. And I don't do monkeys.' So I guess that's off-limits, between the armed guards and the monkeys."

What started as a retirement project has given her a way to share a sense of wonder with other people. "When you look at a pine cone, you might not realize that it took the tree two years to produce that cone. A single cone is a way to look at the whole natural history of pine trees—the way they ward off insects, what their relationship is with fire, the whole ecology of a pine forest is right there. People might just see a pine cone on the ground, but they never stop to think about the story behind it."

"I SPENT A LOT OF MY YOUTH REALLY ANGRY AND WANTING TO DESTROY THINGS, AND I WAS HAPPY TO NOW BE GROWING THINGS."

THE UNAUTHORIZED FORESTER

JOEY SANTORE

Alpine, Texas

IF YOU WANT TO SEE JOEY SANTORE'S TREE COLLECTION, YOU'LL have to look along the median strips of a thoroughfare in his former hometown of Oakland, California. He was so dissatisfied with the municipal planting scheme that he decided to make some improvements of his own. "The city kind of dropped the ball on their public beautification efforts, so I don't think they mind too much," he said. "Most of what they planted has died. The soil is shit quality. You need a pickax in some cases."

For his unauthorized improvements, he planted trees along stretches of Oakland's Mandela Parkway. He chose species that were better suited to the climate, often California natives that he'd grown from seed at his house. "I wasn't growing these trees because I wanted to plant them outside," he said. "I just liked being around them. It's going to sound corny, but they were my friends, you know? I spent a lot of my youth really angry and wanting to destroy things, and I was happy to now be growing things. I just ran out of space in my yard. Then I saw all this empty space, and I was like, well, fuck it, I'm just going to plant it there."

He documented the results of his renegade tree-planting campaign on his popular YouTube channel, Crime Pays But Botany Doesn't. For his videos, Santore (not his real name, and he goes by Tony Santoro online) adopts a heavier version of his native Chicago accent as he narrates plant-hunting trips around the world, with a little social commentary sprinkled in. Although his tattooed hand occasionally makes an appearance as he points to a leaf or a flower, he's not on camera much. The plants are the stars of the show. As a narrator, he comes across as the streetwise, opinionated, vaguely thuggish botany instructor you never knew you needed.

Any tree geek who happened to drive by the Mandela Parkway would be impressed with what he planted there: a Guadalupe Island cypress (*Hesperocyparis guadalupensis*), collected in Baja California and grown from seed, stands alongside a native live oak, *Quercus agrifolia*. ("I grew these bastards from acorns.") Native cypresses and sycamores feature prominently in his unauthorized landscaping, because they're well adapted to the climate and grow quickly. Good criminals know how to blend in: these trees have to get a foothold and establish themselves right away, so they won't look out of place and draw suspicion. A mature tree that looks like it's been around for a while is more likely to be left alone by city maintenance crews.

Of the sixty or so trees he planted along the parkway, more than half are alive and thriving, many of them over thirty feet tall. He considers that a good survival rate for a public, heavily trafficked area. "Some will die," he said. "Be persistent. Keep coming back. Be the case of lice in the kindergarten class—they can't get rid of you."

As a self-taught botanist, he encourages people to figure out gardening for themselves. "The only way to learn is by going out and doing it. You can't learn this shit in a book. Well, you need a good beginning botany book, but that's it. Let your activities be experiments. If you plant something and it dies, find out why. Pull it up. Look at the roots. Was it a fungus? Was it an insect? Figure it out. But you can't just plant it and forget about it."

One benefit of his under-the-radar public tree planting is that it adds native trees to municipal landscapes. "You look at most city plant lists, and they're fucking garbage," he said. "Really tacky, overused trees that have no context, they're not native, and they don't provide habitat. And habitat's getting destroyed by this tumorous economy we have, this cancer-like growth that gobbles up land and turns it into pavement and strip malls and other depressing shit. So you go out and you plant some native trees, and that turns human beings into what they're supposed to be, which is stewards and managers of the land rather than single-use occupants. That's what's rewarding. It gives people a connection to the land they live on."

JOEY SANTORE'S TIPS FOR UNAUTHORIZED FORESTRY

1. Plant native or noninvasive trees suited to your climate.

2. Time it right. Plant when it's cool and rainy.

3. Growth rates matter. Choose a fast-growing tree that can outrun the authorities.

4. Plant a tree when it's a foot or two tall. A slightly larger tree can actually take longer to get established.

5. Guard it against being trampled and pissed on by dogs.

6. Tree companies make good accomplices: ask for a load of free mulch.

7. Return often to the scene of the crime to pull weeds.

8. Don't worry if a tree dies. Plant another one. Keep doing it; you'll figure it out.

"YOU HAVE TO HAVE THE UTMOST RESPECT FOR THE TREES. THEY WILL LET YOU KNOW WHAT THEY NEED."

THE HOLLY COLLECTOR

SUE HUNTER

Felton, Pennsylvania

Sue Hunter feels that she didn't choose hollies—they chose her. "It just seemed to me that the hollies were always there," she said.

She grew up with a family orchard of native American hollies, *Ilex opaca*. "They were enormous—maybe fifty or sixty feet tall. It had a very calming effect on me, to go and sit among the hollies."

As a child she spent a lot of time outdoors, always happy to be by herself with trees as her only companions. "I was very shy as a child. I was always outside in the woods. I learned the names of all the trees very young."

There were other signs that she might devote her life to holly trees. "These connections would pop up, and I didn't really know what they meant at the time, but as you get older and you reflect on your life, it starts to make sense. For instance, I used to collect holly berries and hide them in tins under my bed, until my mother found them, all covered in bugs, and threw them out. And actually, my parents thought about naming me Holly, but they didn't like how it sounded with my last name. Then I played a little holly girl in the school play." She remembers belting out the line "Haul out the holly!" from the stage.

Perhaps it was preordained, then, that she would go to work as a propagator in nurseries and eventually open her own native plant nursery, devoted mostly to the native American holly. Today her nursery sits on about seventy acres, half of which is mature woods, and half of which is cultivated gardens and nursery stock. The grounds are preserved through a land trust and certified as an arboretum through the Holly Society of America.

"I would say I have a thousand holly trees in the ground," she said. "That's over sixty different species. But then I probably have two or three hundred selections I've made. A selection just means that I've chosen a single tree from a naturally occurring population for some outstanding characteristic. Hollies are like people. They're all a little different from each other, just like we are, even though we're all the same species."

These selections can be cloned through propagation, but the offspring they produce from seed won't be exactly like them, just as human children aren't exactly like their parents. "I believe in letting nature do its thing, and if I see an outstanding specimen, I'll select that and take cuttings of it." The selections she chooses might have darker leaves, deep-red or bright-orange fruit, or they might be very fast growing or easier to raise in a container. But to discover these qualities she had to get to know the plants intimately.

"I live, eat, and breathe hollies," she said. "You have to spend years living with them and working with them, and you have to have the utmost respect for the trees. They will let you know what they need."

She's particularly passionate about the native American species *Ilex opaca,* which is considered threatened in her home state of Pennsylvania, and has made it her mission to educate and inspire people to plant them. "You could not ask for a better plant for wildlife. Dozens of species of bird depend on the fruit. The jagged edges of the leaves offer protection for birds when they're nesting, or even around the base of the trunk. They're extremely tenacious. They'll grow on the sides of cliffs, they can take a drought, the wood is durable and strong, and they're evergreen."

Something about the way they persist appeals to her. "They hold on to their leaves all through the fall, and in the spring, they drop the older, inner leaves as they're putting out new growth. That's what I love about them—they don't fit the norm. They do everything in the opposite way. I just want to populate the world with more hollies."

"IF I WAS FINISHED, MY LIFE WOULD BE OVER,
I THINK. I'LL NEVER STOP PLANTING TREES."

THE COLLECTOR'S COLLECTOR
BEN ASKREN
Aurora, Ohio

THE IDEA FOR A TREE MUSEUM OCCURRED TO BEN ASKREN MORE than thirty years ago, when he was a student in a dendrology class. "That class changed my life," he said. "I just started looking at trees, and it was like, wow, look at how different each species is. But the problem was that if you wanted to compare different maples, for example, you had to walk for miles and miles. They were never together. I had the idea way back then that it would be so neat if all the tree families were together, so you could see their similarities."

Throughout his career as an arborist, he worked for tree service companies, consulted with small towns on their urban forestry projects, and always looked for an opportunity to start his museum.

"I got close once," he said. "I was working for a city and they needed a tree plan for an old cemetery. I thought this would be the perfect place to do it. I could run all the trees out in order throughout the cemetery. But when you're working with a city, sometimes the plans change, and you don't know about it. So I had all these trees delivered, and when I got there, they'd been planted already—and not in order!"

But he had another chance in 2014, when the town of Aurora, Ohio, where Askren now works as city arborist, began developing a plot of farmland it had acquired. "The plan was to put in sports fields," he said. "The Audubon Society owned a forest adjacent to the land, and we needed a buffer."

A tree-lined path was exactly what the site called for. It didn't take too much convincing to get permission to turn that path into a tree museum. Askren filled it with the kinds of rarities that tree collectors love and planted them in a sequence that could be used to educate budding dendrologists.

"The idea was to place the trees within their families, in order of how they appeared in the fossil record, but nobody exactly agrees on that," he said. Paleobotanists make new discoveries, and plant taxonomists reassign species to different families, so the sequence Askren had in mind would be subject to revision.

"Everybody agrees on the fossil record from ginkgo to hemlock," he said. "When you get to the flowering trees, it gets weird. Nobody can agree. Used to be one guy would decide something like that, but now it's like NATO. You've got international representatives who get together every ten years and try to decide. I got it as close as I could." He started with ginkgoes (*Ginkgo biloba*), which appeared in the fossil record more than two hundred million years ago, and ended with Russian olive (*Elaeagnus angustifolia*), a relative newcomer that arrived only about twenty million years ago.

Askren loves to walk the path, now lined with four hundred and eighty species comprising forty-four families, to point out the changes that occur as new tree families come on the scene. "You can follow pollination and seed dispersal through the ages. You can see when flowering trees come into it with the magnolias. Pollinators start to appear in the timeline—moths, bees, ants, even animals."

Within this timeline are the kinds of trees that collectors obsess over. He purchased some historic trees from the American Heritage Trees nursery in Tennessee, which sells offspring of trees that grew at the home of a famous person, or that have some connection to a significant person or place. "I've got a Robert Frost birch, a William Faulkner Osage orange, an Edgar Allan Poe hackberry, an Alex Haley pecan, and an Amelia Earhart maple. And speaking of air travel, I got a moon tree descendant from Mississippi State. I still need to make signs explaining what all these are. A museum has signs. I'm trying to make it the same way with trees."

He takes great pleasure in rattling off the interesting and unusual trees in his museum. "I have a Virginia round-leaf birch [*Betula uber*]. It was the first tree to be put on the endangered species list. It's doing better now. I'm also growing forty-eight of the fifty state trees, in order, by year of statehood. I'm only missing Arizona and Hawaii—they just can't survive the winter here."

Many of his trees struggle to make it through Ohio's winters in their early years, and he fears that one harsh season could kill some of them. He keeps a few in six-foot containers and moves them to a greenhouse in the winter, but he can't do that for every young tree. "I know I'm going to have to keep planting replacements," he said, "but that's okay. If I was finished, my life would be over, I think. I'll never stop planting trees."

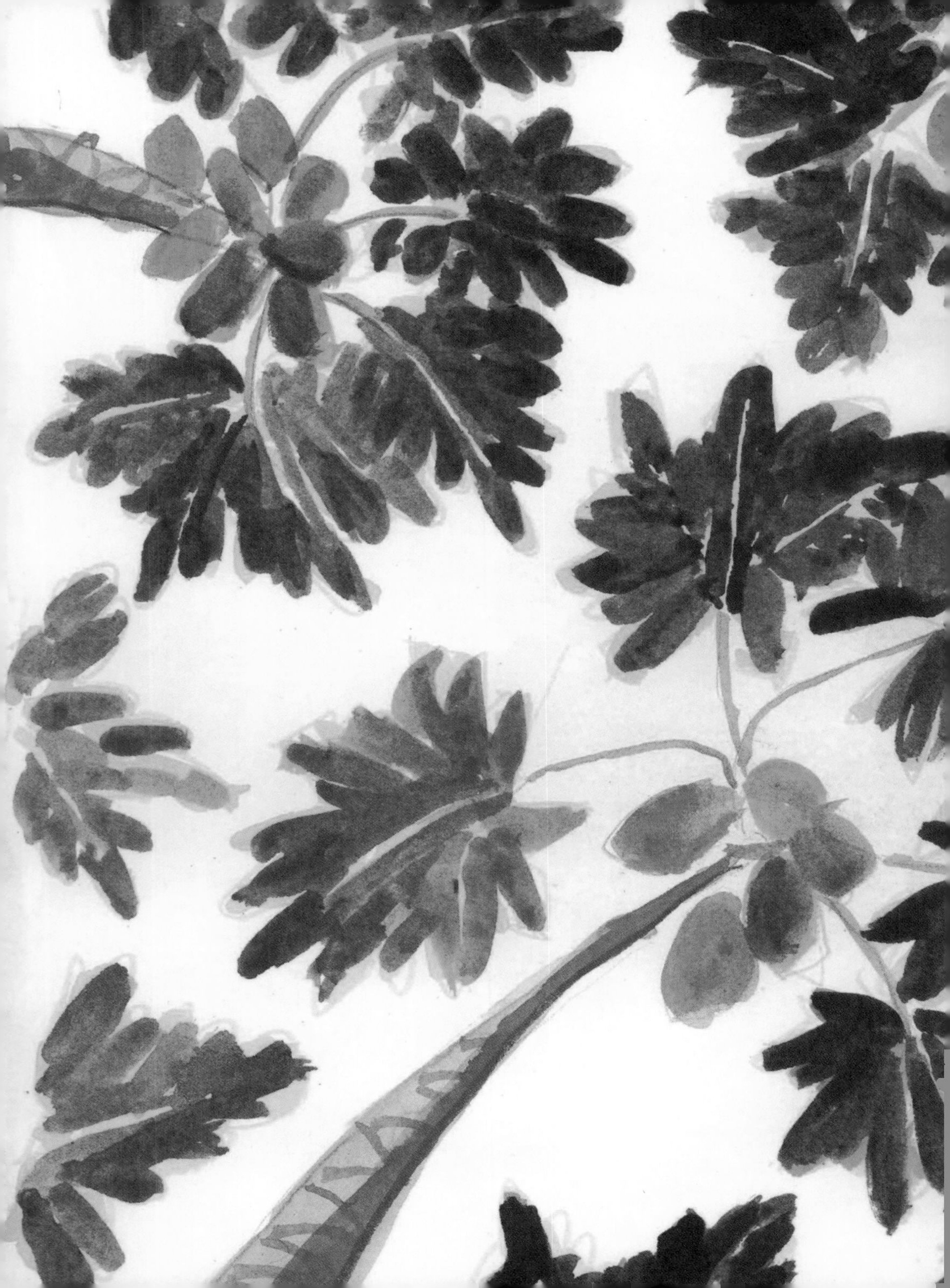

COMMUNITY BUILDERS

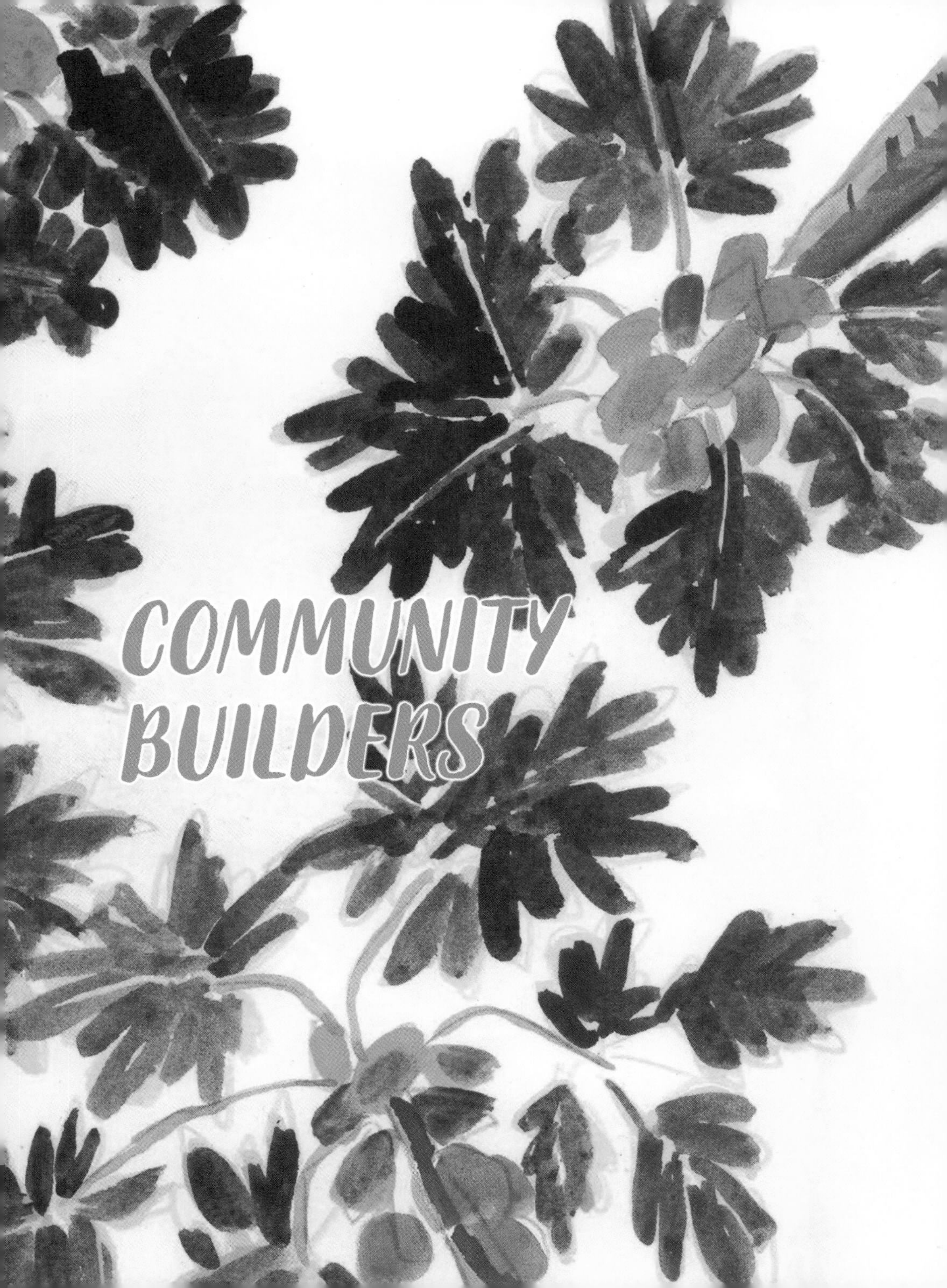

"THEY WERE ONCE PART OF ONE BIG FOREST. THEY CAN
BE CONNECTED AGAIN."

THE FENCE BUILDER

ALEMAYEHU WASSIE ESHETE

Bahir Dar, Ethiopia

IT MIGHT SEEM FAR-FETCHED TO CONSIDER NEARLY EVERY TREE in a state to be part of a collection. But in northern Ethiopia, churches have become the unlikely curators of almost every tree left standing.

The Amhara and Tigray regions, which combined form an area roughly half the size of California, were once densely forested. But three thousand years of agriculture have taken their toll. Over the last century in particular, the human population has dramatically increased, and so has the pressure on the land. Now the terrain is given over to livestock and is almost completely bereft of trees—except around the churches.

"In the Ethiopian Orthodox Tewahedo Church, there is a belief that you come into the church through the Garden of Eden," said Dr. Alemayehu Wassie Eshete, a forest ecologist working in the region. "So every one of these churches is surrounded by a ring of forest. Those are practically all the trees that remain here. If you look at a picture from above, you will see little islands of these forests surrounded by bare land."

Alemayehu remembers the church forests from his own childhood. Their leafy canopies offered a cool, shady respite for churchgoers, but they were also filled with birdsong, the buzzing of insects, and the chatter of monkeys. Everything that remained of Ethiopia's old forests could be found there, around a circular sanctuary covered in a tin roof. Inside the sanctuary, the murals were even painted with the pigments of tree bark, leaves, and fruit. The church was in the forest, and the forest was in the church.

"I went away to university to study forest ecology," he said, "and when I returned, I could see that the forests were shrinking. There's so much pressure to expand grazing land and produce more livestock. The animals trample the church forests and destroy the seedlings. We've already lost ninety percent of our forests, and they're ruining what's left."

Through his work at various nongovernmental organizations over the years, he's found ways to preserve Ethiopia's church forests. "I could not have gone in and told these communities what to do," he said. "The answer had to come from them. The priests found their own solution."

Within each church forest, surrounding the building itself, is a low stone wall meant to mark the entrance to a sacred space. From above, each forest takes the shape of a tire, with the stone wall resembling a narrow rim around the hubcap.

"The priests decided to make another wall around the forest itself," Alemayehu said. "It's not a wall to exclude anyone, but to bring the forest into the church's protection." Local people build the walls themselves, from stone that the farmers are all too happy to have removed from their fields. The walls are high enough to discourage cattle from grazing, but they include a gate to welcome people.

With the walls in place, the trees are able to regenerate on their own. Birds and bees flourish in the forest, and the walls become habitat for insects and lizards. Eighty percent of pollinators in the region live inside the church forests. The trees help to increase the water table, act as a windbreak, and provide much-needed shade.

Alemayehu has cataloged about two hundred different species of trees and woody shrubs that grow around the churches, almost all of them native to the region. An African juniper, *Juniperus procera,* is particularly important to the region's ecology, as is an endangered medici-

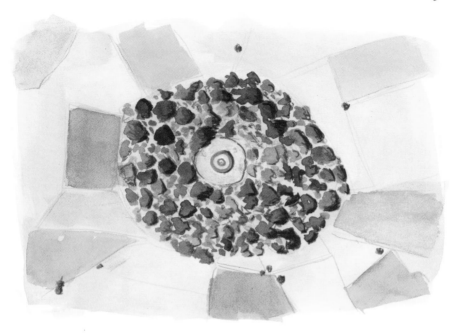

nal species, *Prunus africana*. "These are fragments of ancient forests," he said. "Some of these trees are not particularly rare elsewhere in the world, but we're losing them here."

Priests who take on the construction of the walls are given grants to complete the work and to build latrines, which not only benefit church-goers but also protect the forest ecology. So far, Alemayehu's been able to raise funds through the nonprofit he directs, ORDA Ethiopia, to support the building of walls around thirty-six church forests.

How many more are in need of protection? Only about thirty-five thousand. But Alemayehu is undaunted. He pointed out that the forests are, on average, just two miles from each other. "People think they are too small. Yes, they are small, but many. You can walk from one to the next. They were once part of one big forest. They can be connected again."

"YOU MEAN TO TELL ME
THERE'S A JOB WHERE YOU
GET TO DECIDE WHERE
THE TREES GO?"

THE LANDSCAPE ARCHITECT

DIANE JONES ALLEN

New Orleans, Louisiana

"It wasn't until I was about to graduate college that I found out about landscape architecture," said Diane Jones Allen. "I was just floored. I was like, you mean to tell me there's a job where you get to decide where the trees go?"

Her career as a landscape architect has given her many occasions to decide where the trees go. One of her first opportunities to put her love of trees on display was in her design of a jazz grove in Louis Armstrong Park in New Orleans. "We were planting trees that are used as musical instruments. So I got to research what kind of trees would've been used for drums and other indigenous and African instruments."

A more personal tree collection was born in 2005, after Hurricane Katrina, when she and her husband, Austin Allen—also a landscape architect—were living in the Lower Ninth Ward. The house next door came up for sale, and even though it was in need of major renovations, they bought it with the idea of turning it into a studio space and living quarters for their retirement. But the biggest attraction was the size of the lot.

"Being landscape architects, we had ideas about that land. And we're vegetarians, so we eat a lot of fruit." They'd already been growing fruit trees in pots, and they found that the trees were both easy to grow and easy to share. "You know, you can put an avocado pit in a pot and it will grow. Then we got a banana tree from a neighbor. Now we're trading fruit all over the neighborhood."

More varieties of tropical fruit grow in New Orleans than ever before due to the warmer winters brought about by climate change. Mango and papaya, usually more at home in a Mexican or South American rainforest, now thrive in her neighborhood. The papaya trees in particular are real showstoppers: from a single trunk, like that of a palm tree, hang clusters of enormous fruit ripening from green to orange, sheltered by a canopy of dramatic, deeply lobed leaves. These trees are already dressed for Mardi Gras.

The occasional cold snap will make them die back, but Allen is determined. "I've just got to get them to a certain size and they'll survive," she said. And the occasional freeze isn't all bad: it might even benefit a peach tree she couldn't resist planting. "Peaches actually need a winter chill. We don't get fruit from that tree unless it's been cold for a while."

The vicissitudes of a changing climate will ultimately decide the future of her collection.

Citrus trees are ubiquitous in Louisiana, and Allen grows oranges, lemons, and loquat, along with pomegranates and other pass-along plants from the neighbors, like ginger and grapevines. She's also planted cypress trees, which a local nonprofit was distributing to try to restore the tree canopy in the Lower Ninth Ward. Spruce and pine trees are now making their way into the collection because of her husband's tendency to plant potted Christmas trees outside after they've served their purpose as holiday decor. "He takes environmentalism a little too far," she said. "We're going to have to get more land if this keeps up."

But it's the idea of an extended collection of fruit trees, tended by neighbors up and down the block, that offers a sense of community and the possibility of better harvests for everyone. She and her husband helped to plant an orchard at their church. They gave one neighbor an orange tree seedling, and while theirs didn't make it, his thrived. "We always congratulate him on his green thumb," she said. Another neighbor grows lemons and pomegranates, which they swap for bananas from Allen's garden.

Their avocados depend on good neighbors too: the trees are perfect-flowered, which means that they produce both male and female flowers. But the flowers are never open at the same time, to avoid self-pollination. That's why an avocado tree needs a mate nearby. "I count on my neighbors to grow a few so we'll all have fruit," she said.

Many of the trees are interdependent, relying on pollen in nearby backyards to produce fruit. That sense of community, between both the trees and the neighbors, satisfies Allen's career ambitions as well as her personal passions. "We plant our gardens more as tree lovers," she said, "and we do landscape architecture because we want to make communities better."

"I THOUGHT WE SHOULD PLAN TREES IN THE NAMES OF ALL THE DAUGHTERS WHO ARE BORN IN THE VILLAGE."

THE CHAMPION OF GIRLS
SHYAM SUNDER PALIWAL
Rajasthan, India

IT TOOK THE DEATH OF HIS DAUGHTER FOR SHYAM SUNDER PALI-
wal to realize that something had to change in his village.

"In 2007, my daughter suffered from dehydration. We could not understand what was happening. Immediately she died from low blood
pressure. It was a very painful event that made me understand the value
of a daughter. After twelve days, I planted a tree in the memory of Kiran,
and from that I thought we should plant trees in the names of all the
daughters who are born in the village."

He had good reason for wanting to celebrate daughters in particular.
In 1990, Nobel Prize–winning economist Amartya Sen wrote about the
problem of India's "missing women," citing disparities in birth rates that
pointed to abortion or infanticide of girl children. Government policies
had made some headway in addressing the issue, but Paliwal knew that
girls were still undervalued. "Parents believe that girls are not going to
stay in the family. They have to go to their in-laws' house, whereas the
boy will grow up and earn and make old age pleasant. He would become a support for parents in old age."

As the *sarpanch,* or elected leader, of his village of Piplantri, in the
state of Rajasthan, Paliwal saw an opportunity to make a change. He
proposed a plan to plant 111 trees every time a girl was born. At first
he was only going to plant three trees. "The original idea came when
we wrote down on paper '1 tree for the girl, 1 for her mother, and 1 for
her father.' The digits read like 111 and we decided to plant 111 trees in

the name of the newborn girl. We consider 111 to be an auspicious number."

Since 2007, families have come together to celebrate the birth of girls by planting trees at the outskirts of the village and promising to care for those trees. At a summer festival called Raksha Bandhan, girls tie a string around the trees planted in their honor.

"Historically, girls would tie a string around their brothers' wrists, and it meant that warrior brothers would protect the honor of their sisters," Paliwal said. "In our festival, the trees are planted by the parents, so the girl considers them to be her brothers. But the roles have been reversed because she is swearing an oath to protect them."

Neem trees (*Azadirachta indica*), rosewood (*Dalbergia sissoo*), and banyan (*Ficus benghalensis*) are among the several dozen species that have been grown by local nurseries for this new effort. All this planting has built a flourishing ecosystem, so that trees have started to sprout on their own. Between ceremonial plantings and volunteers, the village is now surrounded by about half a million trees.

This forest is more than a symbol of the value of a girl's life. It provides shade and clean air, a place to picnic, and a habitat for wildlife, and it helps to raise the water table by preventing evaporation and allowing rainwater to percolate into the soil. The village is situated in the shadow of enormous marble mines, which have harmed air and water quality. "We are working on the reparation of that loss," Paliwal said.

What started as a memorial has become an eco-feminist and economic development program that other villages have emulated. Cottage industries have sprung up around some of the plants growing in the forest: in addition to harvesting mangos and other fruit, women cultivate aloe to make juice or gel, and they make bamboo into furniture.

Paliwal paired the forestation project with another idea: families who plant trees for their daughters are also given a savings account, funded by contributions from the villagers, to provide for a girl's education and future dowry. The parents commit to making sure that their daughters go to school and do not marry before the age of eighteen. About a thousand girls have benefited from this program so far.

For his work, the Indian government has awarded Paliwal one of its highest civilian honors, the Padma Shri award. But what matters most is the transformation of his own village. "The last rites of my parents were done here," he said. "My offspring were born here. I want all the villagers to be able to say 'I can do anything for my village. My village is the holiest place for me.'"

"VISITORS ASK US HOW WE MANAGED TO BUY A FOREST ON THE COAST, AND WE TELL THEM THAT WE DIDN'T. WE BOUGHT A DAIRY FARM ON THE COAST, AND WE PLANTED A FOREST."

THE COMMUNAL COLLECTOR

MAX BOURKE

Canberra, Australia

MAX BOURKE'S TREE COLLECTION DOESN'T JUST BELONG TO HIM. It belongs to a group of sixteen longtime friends who banded together to buy a treeless plot of land forty years ago.

Actually, it wasn't entirely treeless. "We did find three *Pinus radiata*, along with some tree stumps," he said. "That was all." Those three trees, commonly known as Monterey pines, weren't even native to Australia. They were from California.

The land, about sixty acres near the village of Central Tilba, had been used as a dairy farm. But photos from the nineteenth century showed that it had once been entirely forested.

Bourke is a plant scientist and conservation expert who was serving as director of the Australian Heritage Commission when he and his friends—including scientists, historians, and teachers—came up with the idea to restore the land, build a vacation house, and enjoy access to about a mile of unspoiled coastline.

"Some of our group were after a nice holiday house, which is fine," said Bourke, "but many of us wanted to restore the forest." They started collecting seed from native trees along the coast. In the last forty years, they've planted about a hundred and fifty thousand trees on the property, many of them the endemic spotted gum, *Corymbia maculata*. Some of those trees are now a hundred feet tall.

"Visitors ask us how we managed to buy a forest on the coast, and we tell them that we didn't," Bourke said. "We bought a dairy farm on the coast, and we planted a forest."

Although the plan from the beginning was to grow only native trees, any rule that governs sixteen friends, along with spouses, children, and guests, is bound to be broken. "One member who recently died at the age of ninety-five snuck in a few oaks, which I was always threatening to chainsaw," Bourke said. "There are also six cottonwoods. I'm going to get rid of those one day too."

But even Bourke found a way to indulge his interest in slightly more exotic trees, by bringing in a native Australian tree from a different part of the country. "I've had a passion for a very strange eucalypt called *Eucalyptus conferruminata,* which is a species that comes from one island off the southwest coast. It produces the most amazing foliage and fruit, unlike other eucalypts. It's a very interesting little tree. I've grown it over the last thirty-five years as a sort of curiosity, I guess. So we put several hundred of those in."

Over the decades, the group has seen births, deaths, marriages, and divorces. Children are inheriting their parents' shares and looking ahead to the next generation of owners. All the while, the trees have filled in and returned the land to what it once was. Only one small area remains bare. "We like to have glimpses of the sea from the house," Bourke said, "so we keep part of it open for the view."

While the forest was growing, Bourke was busy with his family and a career in public service for which he was honored in 2004 with the Order of Australia. But this forest, planted collectively with a group of longtime friends going back decades, brings him particular joy.

"One of my greatest prides is having created this place with this group," he said. "Trees are the reason we're alive on the planet as humans. That seems to me rather important."

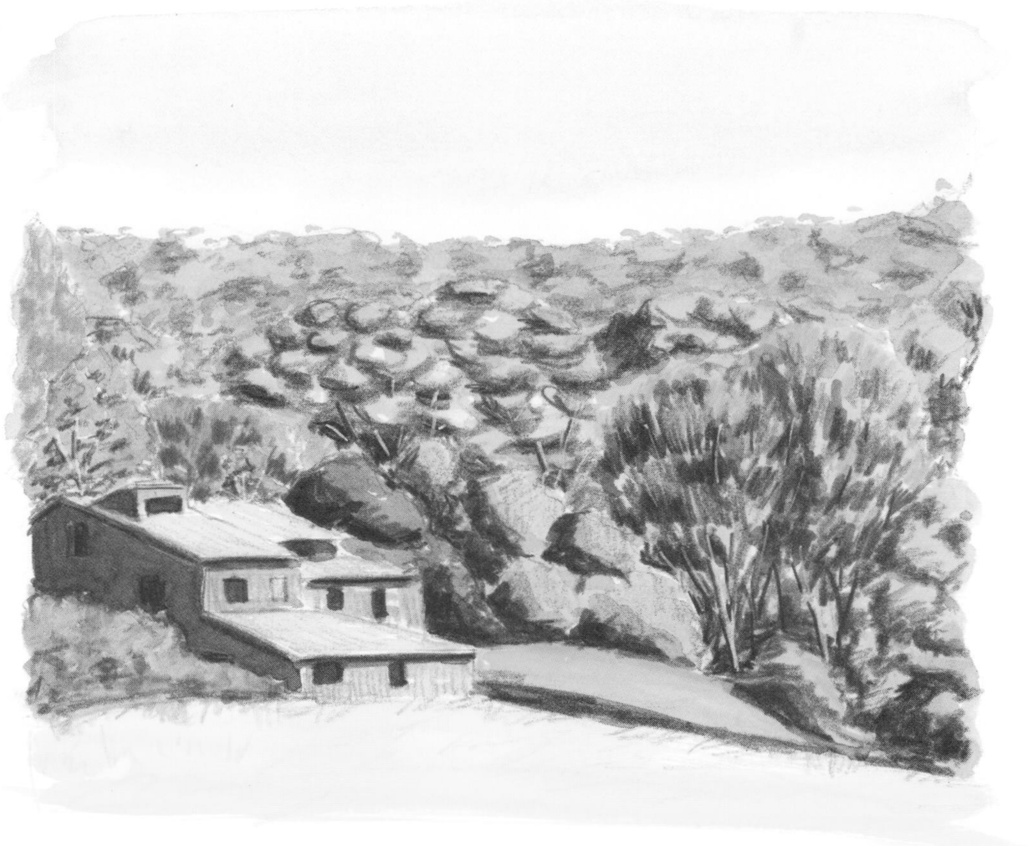

"I'M FINDING OUT THAT I'VE BEEN LIVING AMONG FAMILY I DIDN'T KNOW I HAD."

THE PEACH CARETAKER

REAGAN WYTSALUCY

Monticello, Utah

"I started college as a business major, but I just knew that wasn't what I was supposed to be doing," Reagan Wytsalucy recalled. "So I went to my dad for advice. He's the one who told me about the Navajo peach trees."

Wytsalucy grew up in Gallup, New Mexico, in what she remembers as a very westernized childhood. "One time I heard these words, *Native American*, and I asked my dad what that meant. He just looked at me and said, 'That's you. You're a Native American.' I was ten. I didn't know."

She started to learn about her heritage. In 1864, the United States Army forced Navajo people off their land in New Mexico and Arizona. The army marched them four hundred miles to Fort Sumner, New Mexico, an event that came to be known as the Long Walk. Hundreds of people died along the way, conditions at the internment camp were inhumane, and the army's attempt to relocate the survivors was ultimately a failure. By 1868, those who remained at the camp walked home. They returned to a ravaged landscape. The army had destroyed their agriculture, including their peach orchards.

Captain John Thompson of the First New Mexico Cavalry wrote a report of his efforts to wipe out Navajo peach trees. This report, preserved today at the National Archives, sets out in chilling detail the systematic destruction of peach orchards through the Canyon de Chelly region of eastern Arizona. It began on July 31, 1864, when "I found there in a Peach Orchard Containing about 200 Fruit Trees all bearing Fruit, which I had Cut down." A few days later, he wrote, "I cut down 500 of the best Peach trees I have ever seen in the Country, every one of them bearing Fruit." In all, he documented the eradication of forty-five hundred fruit trees. Others came along after him and took down still more orchards.

Those peach orchards were a vital part of Navajo life, and provided a valuable commodity for trading. Toppling those trees wasn't meant only to wipe out a source of wealth for Navajo people, however. It was also meant to starve out anyone who had escaped the Long Walk by hiding in the canyon. One of those who hid and survived was Chief Hoskinini, Wytsalucy's ancestor, who lived until 1912.

"He was a caretaker," she said. "He rounded up livestock that was left behind. When people came back, he had sheep for them, and cows, so they could start again." He also would have known where the surviving peach trees could be found, so the seeds could be planted and the orchards restored. Her father remembered eating those peaches as a child.

"My dad wasn't sure if any of the trees could be found," she said. "But it inspired me to get a bachelor's degree and learn everything I could about horticulture." When she told a professor about the Navajo peach trees, he encouraged her to pursue it as a master's thesis project. "He said, 'You can research this crop, and look for them, and we'll get grant funding,' and I was like, 'Wait, you're going to pay me to do this?'"

It was two years from the beginning of her project until the day she had peach tree seeds in her hands. She visited farms and remote sites of old orchards, met with tribal leaders, sought approvals from every community where the trees might be found, and developed a program to grow the trees and safeguard them as a resource to benefit the tribes. The first few dozen of these trees are growing in a nursery in Utah, and she has plans to germinate still more seeds.

Navajo peaches tend to be small, white-fleshed, and well suited to being dried and preserved. Because Navajos never grafted their trees, and only grew them from seed, Wytsalucy is doing the same. Over time she hopes both to identify nursery partners to expand production, and to find a culturally appropriate way to reintroduce the trees. Right now, she continues to search for old orchards—and their caretakers.

"I keep finding people who remember what Chief Hoskinini did for their family. They have these stories about him. So I'm finding out that I've been living among family I didn't know I had. Wherever I go, people open up their homes and offer me a place to stay. Through this work, I'm finding out who I am. I can't explain it any other way except to say that I'm on a divine pathway."

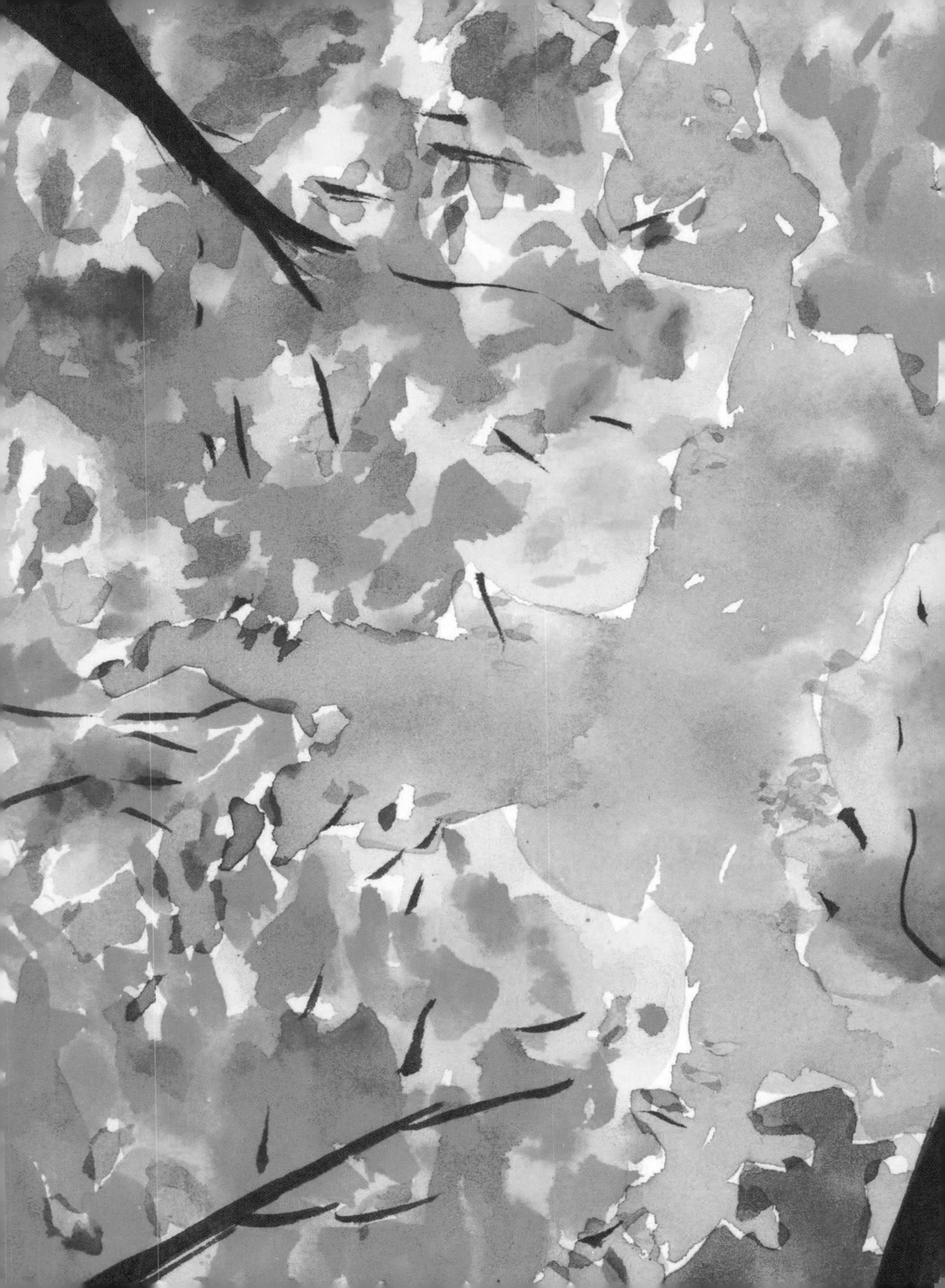

ENTHUSIASTS

"THAT'S JUST MAKING A VIRTUE OUT OF PASSION."

THE HIGHEST BIDDER

SARA MALONE

Petaluma, California

SARA MALONE HAS DEVELOPED SOMETHING OF A REPUTATION AT rare tree auctions. "I'm known as one of the big spenders. So they run around and try to find me as many glasses of wine as they can, like, 'Let's get a drink in that woman.' You never know what I'll do."

Her passion for trees was sparked on a garden tour of England in 1995. "When I look back to the photographs I took, I see the mixed borders, I see the perennials, but so many of my pictures were of the huge trees. What I brought back was a love of these trees, the bark, the texture, the structure, the majesty. That was just a heartfelt response. And then I tried to come up with all these reasons why trees are good, you know, for shade, or cooling, or transpiration or whatever. But that's just making a virtue out of passion."

Tree auctions, which are usually held by botanical gardens and tree societies as fundraisers, satisfy both her virtuous and her passionate sides, as the free-spending collectors are supporting a worthy cause. Malone, who cultivates four acres in Petaluma, California, realized that a major benefit to joining a botanical society was to get access to rare and unusual plants. "So many of these trees at auction are just not for sale in the trade," she said. "Either the grower never bothered to register them, or even if they did, there just wasn't enough of a market for one reason or another. Like *Cladrastis kentukea,* the yellowwood tree. It takes ten to fifteen years to bloom. You just can't get a nursery to carry it, because it doesn't look like anything when it's not blooming. But I'll buy one of those."

The bidding frenzy at a rare tree auction rivals that of an art auction, although a tree fetches far less than a Picasso. "One time, maybe ten years ago, some gal got into a bidding war over a mugo pine [*Pinus mugo*], of all things," Malone said. "That's a pretty common tree. She ended up paying four thousand dollars. There's always this kind of judgmental reaction, like, 'She paid that much for a mugo pine?' I mean, clearly she wanted it. And it's supporting the plant society. But I always wondered what happened to that pine."

Another appeal of a tree auction is that oversized trees are for sale, and some collectors might find it worthwhile to pay extra to plant a significantly more mature specimen. "Trees go through a very long adolescence before they ever start to look like anything," she said. "If you're only ever planting small trees, your place is going to look like a miniature golf course for quite a while. I do think that having a tree collection is possibly different than having a collection of rare books or jewels, or whatever it might be, because the trees have this other dimension of time."

Many of the most intriguing tree auctions take place in Oregon, where the ornamental tree nursery business is concentrated. Malone will drive up from northern California when there's a good auction on. She takes a horse trailer, which is perfect for trees, because it has windows for air circulation and a water tank. "When I come through the ag station going back into California, the inspectors say, 'You got livestock back there?' And I say no, and show them my manifest with all the trees listed, and they just look inside and shake their heads."

The real challenge, though, is getting trees home from far-flung auctions. "In North Carolina, I bought an Afghani pine. Tony Avent from Plant Delights Nursery happened to be the auctioneer. I started bidding on it, then I realized it was six feet tall, and I couldn't get it home. So I hollered out that if I won it, he'd have to ship it to me, and he foolishly said yes. He ended up having to make a huge wooden box to get it to California."

After twenty-five years on her property in California's wine country, she's running out of places to plant those extravagant purchases. More than two thousand different species and subspecies flourish in a landscape that is mature and richly textured, if a little crowded. "My friend Ryan Guillou, who is the curator of the San Francisco Botanical Garden, always tells me, 'Don't ask yourself if it'll live. Ask yourself if it will thrive. And if it will not thrive, get rid of it.' So that's been one of my real editing challenges, because I look at the tree and I go, 'But I love this tree, this tree is so cool.' But mine looks like shit. So that's what I'm doing now, editing out the things that just don't thrive."

"IT LOOKS A BIT LIKE A LAUNCHPAD IN OUR BACKYARD."

THE ZONE PUSHER
DAVE ADAMS
Boise, Idaho

DAVE ADAMS HAD NEVER COLLECTED ANYTHING, NOR HAD HE ever planted a garden, when his husband, John Watkins, brought home a sago palm (*Cycas revoluta*).

"It was just a little golf-ball-sized plant," Adams said, "but I thought it looked sort of interesting, so I took it over. I started reading up on sagos, mostly so I'd know how not to kill it. Pretty soon I thought, *Well, this one didn't die, so maybe I'll try another one.*"

A sago palm is not technically a palm tree—it's a cycad, a type of ancient gymnosperm that can trace its ancestry back 320 million years. But it is a kind of a gateway drug into palm collecting: it's small enough to grow as a houseplant, and its spiky, architectural leaves make it a rather dramatic piece of decoration.

"I started buying palms from a dealer in San Diego," Adams said. "They didn't all make it at first. I had a triangle palm [*Dypsis decaryi*] and a Bismarck palm [*Bismarckia nobilis*] that both died. I would grow those again if I lived in Florida or California."

But he lives in Boise, Idaho, which gets a foot or two of snow in the winter.

"I do a lot of zone pushing," he said. "If the palms are going to survive the winter in Boise, I've got to try to mimic the bare necessities that they need to get by."

A few of the tougher species, such as a Mexican fan palm (*Washingtonia robusta*) and a windmill palm (*Trachycarpus fortunei*), can tolerate cold winters, but even they get some protection. His husband described the process: "To get the outdoor palms through the winter, we thin out the fronds and wrap them in burlap. We then wrap Christmas lights—the large-bulb type—around them to provide warmth. The lights have a thermocouple to turn them on when the temperature drops to thirty-five degrees. Then we wrap them with an insulating cover that has a shiny outside surface. It looks a bit like a launchpad in our backyard. It takes one of us on the ladder and one stabilizing the base of the cover to lower these over the palms."

Only half a dozen palms spend the winter outdoors under their insulation and Christmas lights, while the others come inside. Adams and Watkins live in a suburban home with a modestly sized backyard, so there's a great deal of shifting around to accommodate the palm trees' move to their winter quarters.

"I have a bottle palm [*Hyophorbe lagenicaulis*] that overwinters in our guest room," Adams said. "I have to cut off a lot of fronds just to get it down the hallway. In the winter it looks like our house is a hospice for palms."

The rest hunker down in the garage under grow lights. "They're always going to be in containers because I have to be able to move them indoors for the winter. Every year they get heavier to move. Just pushing them onto a dolly so I can wheel them around is a chore. Some of them are so tall that they're hitting the ceiling of the garage now."

In spring, the palms move outdoors gradually. "I bring them out into the driveway for a few hours at a time, so they don't get scorched. The neighbors are always so happy to see them. When the palms are out, it means that summer is on the way."

ZONE PUSHING: The sometimes foolhardy act of attempting to grow plants that are not normally able to survive in your USDA plant hardiness zone.

COLLECTIVE NOUNS FOR TREES

Some collective nouns are literary and fanciful. "A murder of crows" is an artful way to describe a group of crows, but it isn't necessarily scientific. You might not hear a botanist refer to "a parliament of oaks," "a canvas of maples," or "a convocation of redwoods," but there are, in fact, technical terms for groupings of trees.

Let's begin with some arboreal Latin: An arboretum is a collection of trees. Tree collectors use the Latin *-etum* suffix—defined as a collection of trees or plants—to describe any number of specialized tree collections. Some are obvious—a palmetum is a collection of palm trees—but if you don't know the Latin name for maple trees, for instance, you might not recognize that an aceretum is a collection of maples.

ACERETUM
A collection of maples, derived from the genus name, Acer

BETULETUM
A collection of birches, derived from the genus name, Betula

CITRETUM
A citrus orchard

CONIFERETUM
A collection of conifers

LAURELETUM
A collection of laurels

OLIVETUM
A collection of olive trees

PALMETUM
A collection of palm trees

PINETUM

A collection of pine trees

QUERCETUM

A collection of oak trees, derived from the genus name, Quercus

SALICETUM

A collection of willows, derived from the genus name, Salix

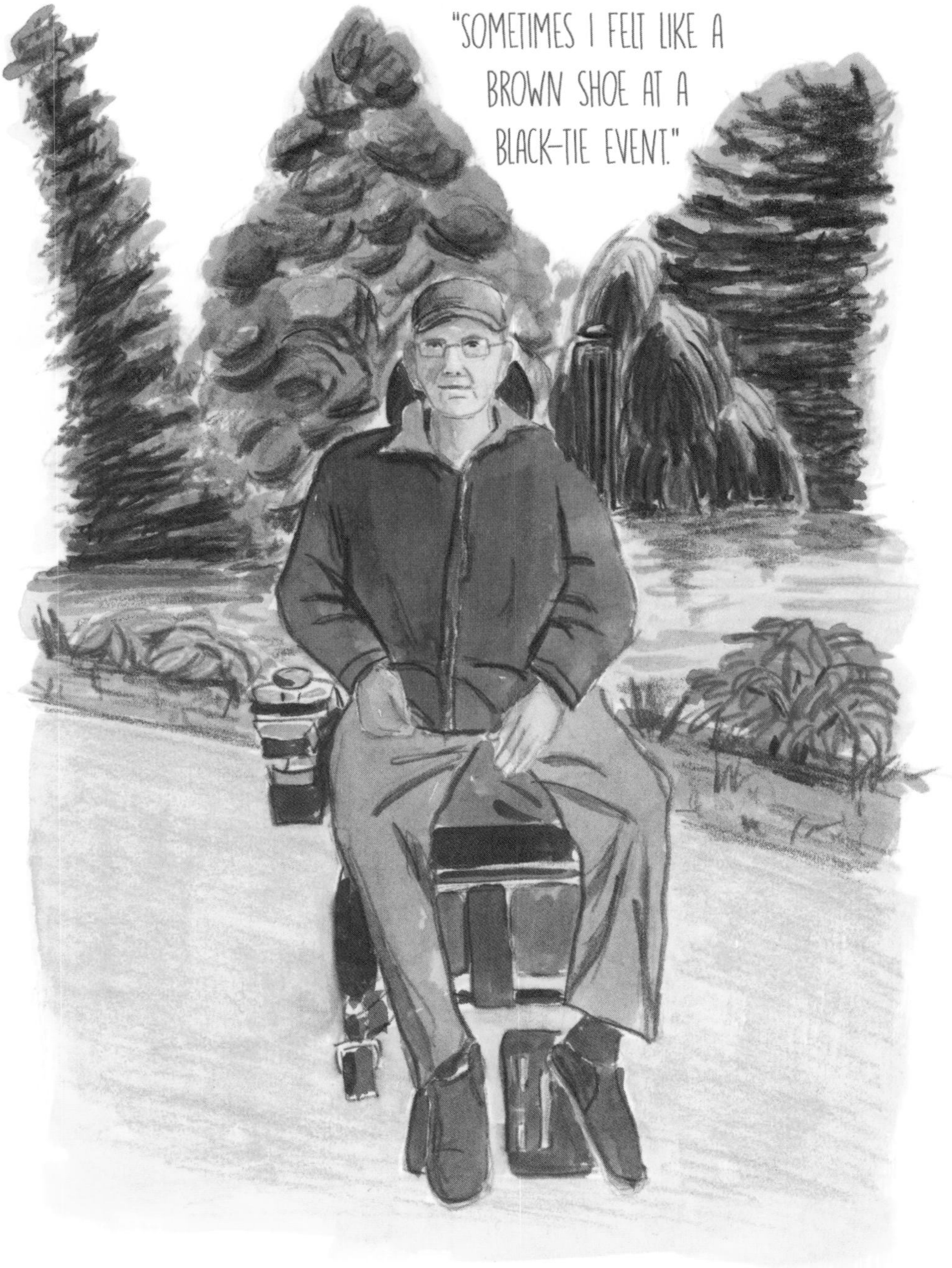
"SOMETIMES I FELT LIKE A
BROWN SHOE AT A
BLACK-TIE EVENT."

THE NOAH'S ARK OF PLANTS

TOM COX

Canton, Georgia

THE ORIGINS OF TOM COX'S PRIVATE ARBORETUM CAN BE TRACED back to his army days, when he'd sneak plants onto the military base to spruce up neglected yards. "It was definitely a situation where it was better to ask forgiveness than permission," he said. "I made the place look good. It was never my property, though."

After twenty years in the army, he retired from military service and took a job with the phone company. "When you look at this tree collection, I want to point out that I did all of this on a phone company salary," he said. "We were never rich." In 1986, he and his wife bought thirteen acres of land north of Atlanta. He had an idea that he wanted to create a Noah's Ark of plants, and he chose that parcel deliberately because it contained a range of Georgia's growing conditions: swamp wetland, upland, and dry shade. "I wanted to grow all the most rare and unusual trees from around the world that could possibly thrive in this region," he said.

He started attending horticultural conferences and meetings of local plant societies. Gradually, a world of like-minded collectors opened up to him. "Sometimes I felt like a brown shoe at a black-tie event," he said. "But people would take me under their wings, these botanists and scientists and plant hunters. They saw that I had an interest and a passion. The Latin names came easily to me. I have an encyclopedic memory for those sorts of things."

Through his contacts at tree societies and other horticultural groups, he and his wife, Evelyn, began to travel and seek out rare plants. "We've gone everywhere, from the Azores to Australia and New Zealand, to Japan and China and cloud forests in Mexico. And I kind of sit back sometimes and pinch myself. Here I am, no special academic background, just a love and curiosity for plants, and now all of a sudden we're accepted into this really exclusive world. It's been a journey."

Today, Cox Arboretum and Gardens is one of the most significant private arboreta in the country. The focus of the collection is primarily on conifers, but Cox grows a little of everything, from rhododendrons to camellias to maples. Nurseries send him new varieties to try before they release them into the marketplace, just to see how they'll do in his climate. "Some of these plants are going to be patented," he said. "They come with a contract that says I won't share any seeds or cuttings."

One of his most prized trees is the Baishanzu fir (*Abies beshanzuensis*), which is perilously close to extinction. Only three are growing in the wild, on a single mountaintop in the southern Zhejiang province of eastern China. (There were seven at one time. One died in the wild. Three were dug up and moved to the Beijing Botanical Garden, where they also died.) "What I have is a specimen grown from a cutting," Cox said. "A collector in England sent one to a nursery in Oregon so they could graft it for me." Propagating a rare tree and introducing it into botanical gardens and private collections is one way to help keep it alive when it's threatened in the wild.

Another tree that is not endangered but vulnerable, the Chinese coffin tree, *Taiwania cryptomerioides,* has been growing in his garden for fifteen years. Wild populations have nearly been destroyed because the tall, straight trunks make the tree extremely attractive as timber for furniture and, among other things, coffins. It's been grown in botanical gardens and in the private gardens of conifer enthusiasts for more than a century. "It's so graceful, with these long, arching branches," Cox said.

"It's a perfect tree in terms of its symmetry. Luckily, I put it in a spot where I can look out a big picture window and see it every day."

That picture window matters more and more. Cox has a rare neuro-muscular disease that keeps him in a wheelchair. But that has only inspired a new project: he's having some of his most prized trees moved near the house, where they're accessible along a paved walkway. "I'm calling it 'the rarest of the rare,'" he said. "I have to move things closer to where I can see them. I have a saying that everything here is on skates, meaning I can move them, and I do move a lot of things. Putting this new collection together keeps me going."

"WHEN I'M WITH THE PEOPLE IN THE MAGNOLIA SOCIETY, I'M JUST ANOTHER PERSON. I'M NORMAL."

THE MAGNOLIA COLLECTOR

BETH EDWARD

Syracuse, New York

BETH EDWARD HAD ALWAYS BEEN A GARDENER, BUT A BOOK about magnolias turned her into a collector. "I used to belong to a garden book club, and one of the selections was a book by Dorothy Callaway called *The World of Magnolias*. At the time, I thought that there were three magnolias: the star magnolia, the saucer magnolia, and the southern magnolia. It turns out there's hundreds of species, and countless hybrids. I had no clue. In the book she mentioned the Magnolia Society. I thought it would be interesting to join, but I didn't really do anything about it."

She was inspired enough, though, to order a few magnolia trees from a specialty nursery. When her order arrived, it included a leaflet that once again mentioned joining the Magnolia Society. That seemed like fate to her, so she signed up and started going to meetings. What she couldn't have predicted at the time was that this group of magnolia lovers would not only teach her about the trees, but give her a new sense of kinship.

Magnolias intrigued her because they're such primeval plants, dating back to the age of dinosaurs. They evolved before bees and relied, instead, on beetles for pollination. Today some are threatened in their wild habitats in Southeast Asia, South America, and the West Indies, and the Magnolia Society holds fundraisers to support conservation efforts. "That's what's so heartbreaking about losing magnolias in the wild," Edward said. "We're killing trees that have been around since day one, basically."

Magnolias aren't just ancient and unusual: they're beautiful too, with their enormous, glossy leaves and luxuriant flowers that bloom over a long season. The first magnolias to open in Edward's garden start so early that they're at risk of freezing in a spring frost. Other varieties bloom throughout the summer and even rebloom later in the year. Her favorite is *Magnolia fraseri,* a North American native with flowers that smell intensely like a piña colada. "It'll scent the whole area around the tree," she said. "It's not endangered or rare, but it's a tree that nobody grows because they don't know about it. I have three of them."

The Magnolia Society opened her eyes to the variety of magnolias that could grow in Syracuse's cold climate. Another member showed her that the trees don't need sun after their leaves drop in the fall. She realized that she could grow more trees in small pots and move them to the garage in winter. "I keep it heated to just above freezing. My husband has these special cars in the garage, and tucked among his special cars are all my trees. Then they come back out in the spring."

Some of those trees were started from seed that came from the Magnolia Society's fundraisers, and they're destined to be planted outdoors once they've had five years or so to mature. Some aren't hardy enough to ever survive a winter in upstate New York, but she can't resist growing them anyway. "I'll keep those in containers until they're too big to move in and out of the garage anymore, and then they'll go to the compost pile," Edward said. "And I'll start again with the next one."

She estimates that she has about fifty magnolias in containers and another hundred in the ground, although that number is constantly changing. "One time I was traveling and my husband counted how many containers he had to water while I was gone. There were sixty-five. He wasn't complaining, he was just reporting. He's an engineer, so he likes data."

The pleasure she takes from her collection and the rewards of her membership in the Magnolia Society are intertwined. The group organizes field trips to arboreta and public gardens, which have given her a chance to travel the world with like-minded collectors. "It really adds a lot to my life when I visit gardens with all these experts from great arboretums, who really know their stuff. You get to walk around and ask questions and hear what they have to say. It's a huge learning experience."

Beyond the trees, though, is the sense of kinship. "A community develops when you're interested in one plant. I work as a computer programmer, and no one else that I know in my career, or even in my family, really shares this interest of mine. I'm the only one. But when I'm with the people in the Magnolia Society, I'm just another person. I'm normal."

A SOCIETY FOR EVERY TREE

If two or more people have ever been interested in a particular tree, you can be sure that they have formed a society to foster its cultivation, appreciation, and conservation. Many of these groups have international, national, and regional chapters. In some cases, a regional group, like California Rare Fruit Growers, operates on an international scale in spite of its name.

**EUROPEAN BOXWOOD
AND TOPIARY SOCIETY**

**INTERNATIONAL
CAMELLIA SOCIETY**

**AMERICAN CONIFER
SOCIETY**

**INTERNATIONAL
DENDROLOGY SOCIETY**

HOLLY SOCIETY OF AMERICA

THE MAPLE SOCIETY

MAGNOLIA SOCIETY INTERNATIONAL

INTERNATIONAL OAK SOCIETY

INTERNATIONAL PALM SOCIETY

CALIFORNIA RARE FRUIT GROWERS

AMERICAN RHODODENDRON SOCIETY

"AS SOON AS I PICK A FAVORITE, I'M BETRAYING THE OTHER ONES."

THE LIFELONG COLLECTOR

LEN EISERER

Lancaster, Pennsylvania

LEN EISERER HAS BEEN A COLLECTOR HIS WHOLE LIFE. "THE VERY first thing I collected was numbers," he said. "I think I was in seventh grade. I just started writing down every number in order. I filled notebooks with them. Then one day, I told the father of one of my friends what I was doing. He just looked at me and said, 'But Lenny, you'll never be finished.' Just him saying those few words—that ended it. It was so clear to me that there was no point to it."

He went on to collect dimes. Not rare or unusual dimes—just any dime he came across. "It's a good coin, the dime. When you have hundreds of them, they feel good in your hands. You couldn't call it a coin collection, exactly, I just had a lot of them."

Then he went back to numbers. "I started noticing license plates. I spent hours out in the church parking lot, or at the grocery store, writing down license plate numbers."

In graduate school for psychology, he collected references, creating lengthy annotated bibliographies on specific topics. "My collections—they go on and on," he said. "I wrote a book about robins, and for a while I collected anything having to do with robins. We'd go to a diner, and they'd have sugar packets with pictures of birds on them. I had to go through them and pull out all the robins. I probably have five hundred of those sugar packets, just because they have a robin on them."

But he hadn't given much thought to trees until 1979, when he and his wife bought a house surrounded by an expanse of flat, monotonous lawn. He wanted to plant some trees along the driveway, so one weekend, he walked into the woods behind his property and pulled out some seedlings. "I had no idea what I was doing," he said. "The trees were Norway maples. They're really invasive. Everyone hates them. But I didn't know any of that yet. Those were my gateway trees." He planted sixteen, and was astonished to watch them reach three to four feet by spring. That made him want to plant more.

"I had no vision when I started," he said, "but I'm very competitive, so when I read about a local farmer who had, like, ninety different trees, I thought, *Well, I just have to have more than he does.*"

His selections were haphazard at first. "I just went kind of nuts for a while. I didn't want more than one of the same tree. Everything had to be a different species or cultivar." It didn't occur to him to design a naturalistic landscape across the two and a half acres he intended to plant. Instead, everything went into rows, like books on a bookshelf.

"I would plant a tree, pace off six or eight paces, and plant another tree," he said. "I could keep track of them that way. I had a spreadsheet." For the first few years of a tree's life, he would measure and record its growth, until it got too tall to measure without a ladder.

He planted about a hundred and fifty trees, and he's tucked even more away in containers around the house. "I even collected the shovels," he said. "I've gone through a lot of shovels planting these trees, and I still have every one."

He hesitates to name a favorite tree. "People ask me that all the time, but as soon as I pick a favorite, I'm betraying the other ones. If I'm being honest, I always root for the underdog. I was bullied when I was young, and it makes me root for the one that gets picked on. Even if a tree is sick or dying, I can't bear to cut it down. I want it to have a chance."

When he ran out of space, he found a way to collect trees virtually. He started a website, Tree Treasures of Lancaster County, to document the most noteworthy trees in his county. "They're not just the biggest trees, or the rarest," he said. "It might be something to do with the history of the tree, or its shape. Any reason people have for thinking a tree is special, they can nominate it."

Collecting with his camera doesn't quell the urge to plant more trees, however. "Anytime I see a tree I don't have, I still kind of lust after it."

"I WAS SITTING UP WITH A SICK TREE."

THE CHARACTER ACTOR
EDWARD EVERETT HORTON
Encino, California

MOVIE BUFFS REMEMBER EDWARD EVERETT HORTON FOR HIS roles in films like *Arsenic and Old Lace* and *The Gay Divorcee,* where he played opposite Fred Astaire and Ginger Rogers. A certain generation will recognize his voice on *The Adventures of Rocky and Bullwinkle and Friends.* But longtime residents of Encino, California, know him as the owner of Belleigh Acres, a tree-lined estate where the actor spent his time entertaining celebrity friends and unwinding in the garden.

Horton was a comedic actor, and what little is known about his tree collection comes from his quips to the Hollywood press. In 1937, a newspaper reported that he was "late for his chore of acting at a studio" and an assistant had to call him at home. Horton apologized for oversleeping. "I was sitting up with a sick tree," he explained. "It is a very expensive tree, a linden I imported from Germany. The tree doctors couldn't do anything for it. I sat up all night, keeping it warm with salamanders, occasionally spraying it, picking off a dead leaf here and there. It began to look healthier when I went to bed—at sun-up—this morning."

He wrote a lighthearted essay the same year in which he talked about having a reputation for being a soft touch. "Take my fan mail to prove it," he wrote. "Not a day passes when I'm not invited to pay off somebody's mortgage." He admitted to being a pushover, claiming to employ enough workers on his estate to solve the unemployment problem in Encino. Even trees took advantage of him: "I have a hobby of buying old trees and transplanting them on my estate. I aim to have a tree or shrub from every country on earth before I get through. When I returned from England, a few weeks ago, I brought several varieties with me. A recent inventory revealed the fact that I have more than four hundred and fifty kinds. Now do you believe me?"

Horton's archives have been carefully preserved, but they yield no clues about the specific trees he grew or what sparked his interest in arboriculture. We do know that his estate included rose gardens, a fruit tree orchard, and a generously proportioned country home to which, according to reporters, he added a room every time he made a picture. Old photographs of the place show endless construction projects and rows of young trees planted in freshly excavated earth.

He claimed to have purchased the property in 1925 from "a bootlegger who was en route to jail." Over time, he said, "I developed a baronial

complex" and expanded to twenty-two acres. In addition to using his property for parties and tree collecting, he offered F. Scott Fitzgerald the use of a cottage there in 1940, while the author was struggling to finish his novel *The Last Tycoon*.

The estate's days as a quiet country retreat ended in 1959, when Horton reluctantly agreed to sell eleven acres of the property, which had to be cleared to make way for the Ventura Freeway. His house was situated uncomfortably close to the new road: in fact, he negotiated for a kind of abutment to be built alongside the freeway to support his back patio. A 1960 photograph shows him sitting in a wicker chair with a chain-link fence and the freeway behind him, his fingers in his ears in protest of the noise. That abutment can still be seen along the freeway today, but the only other reminder of his long-gone estate is the street named after him, Edward E. Horton Lane.

Horton lived at the estate until his death in 1970. In an interview with an upstate New York paper, he once deflected the perennial question about why he never married (his longtime companion and fellow actor Gavin Gordon could've supplied the answer) by changing the subject to trees. On the topic of marriage he said vaguely that he might, someday. "These things are so unpredictable, and happen so suddenly, that you never can tell," he mused, before adding that he looked forward to visiting his family at Lake George, where he liked to sit and watch the birds in the trees and look out over the water. "The quiet days come so rarely in the theatrical life," he said.

HALL OF FAME: CELEBRITY TREE COLLECTORS

JUDI DENCH

Famed actress Judi Dench plants trees on her property in Surrey, England, to memorialize loved ones she has lost. Dench has also championed the preservation of heritage trees threatened by development and supported conservation causes. She's shared her love of trees in other, unexpected ways: when an enormous oak-tree branch fell on her property in 1987, she had a local wood-carver make hearts of the wood and gave them to fellow members of the cast of *Hamlet*. In a documentary about her love for trees, she said, "My life now is just trees. Trees and champagne."

CHUCK LEAVELL

Leavell made a name for himself as the keyboardist for the Allman Brothers Band and the Rolling Stones, but he's also known as a conservationist and tree farmer. He and his wife, Rose Lane, practice sustainable forestry at their Georgia property, Charlane Woodlands and Preserve, and he hosted the PBS series *America's Forests*. He's often said that he has three passions: "My family, my trees, and my music."

JASON MRAZ

The singer-songwriter bought an avocado orchard near San Diego and created Mraz Family Farms. He grows forty fruit varieties and eleven different types of coffee. On the farm's website he's described this way: "Some people collect cars. Jason collects fruit trees." He also plants trees in cities he visits on tour and honored his grandfather with the song "The Man Who Planted Trees."

ALFONSO OSSORIO

Ossorio was that rare abstract expressionist who was not a starving artist: his family made its fortune in the Philippine sugar mill industry, which allowed him to purchase a fifty-seven-acre estate in East Hampton called "The Creeks" in 1952. He befriended, and collected paintings by, almost every major artist working at the time, including Jackson Pollock, Clyfford Still, and Louise Nevelson. He also collected trees, developing one of the most significant conifer collections in the country and welcoming members of tree societies to his garden for tours.

SEBASTIÃO SALGADO

After photographing genocides in Rwanda, the famous photographer was exhausted, devastated, and unable to work. His wife, Lélia, suggested that they return to his family's farm in Brazil. When they arrived, they were astonished to see that the jungle that Salgado remembered from his childhood had been destroyed. For twenty years they oversaw the restoration of the fifteen-hundred-acre property and the planting of two million trees. Today Salgado's nonprofit Instituto Terra is charged with conservation, education, and advocacy.

SEO TAIJI

In 2010, fans of Korean pop star Seo Taiji decided to celebrate the twentieth anniversary of the release of his first album by raising money to plant a forest in his honor. With the help of the World Land Trust, they created the Seo Taiji Forest on former farmland in a critically endangered area of Brazil. Fans funded the planting of five thousand trees to restore the land. Five years later, the star thanked his fans by planting a twin forest nearby honoring them, the Seo Taiji Mania Forest.

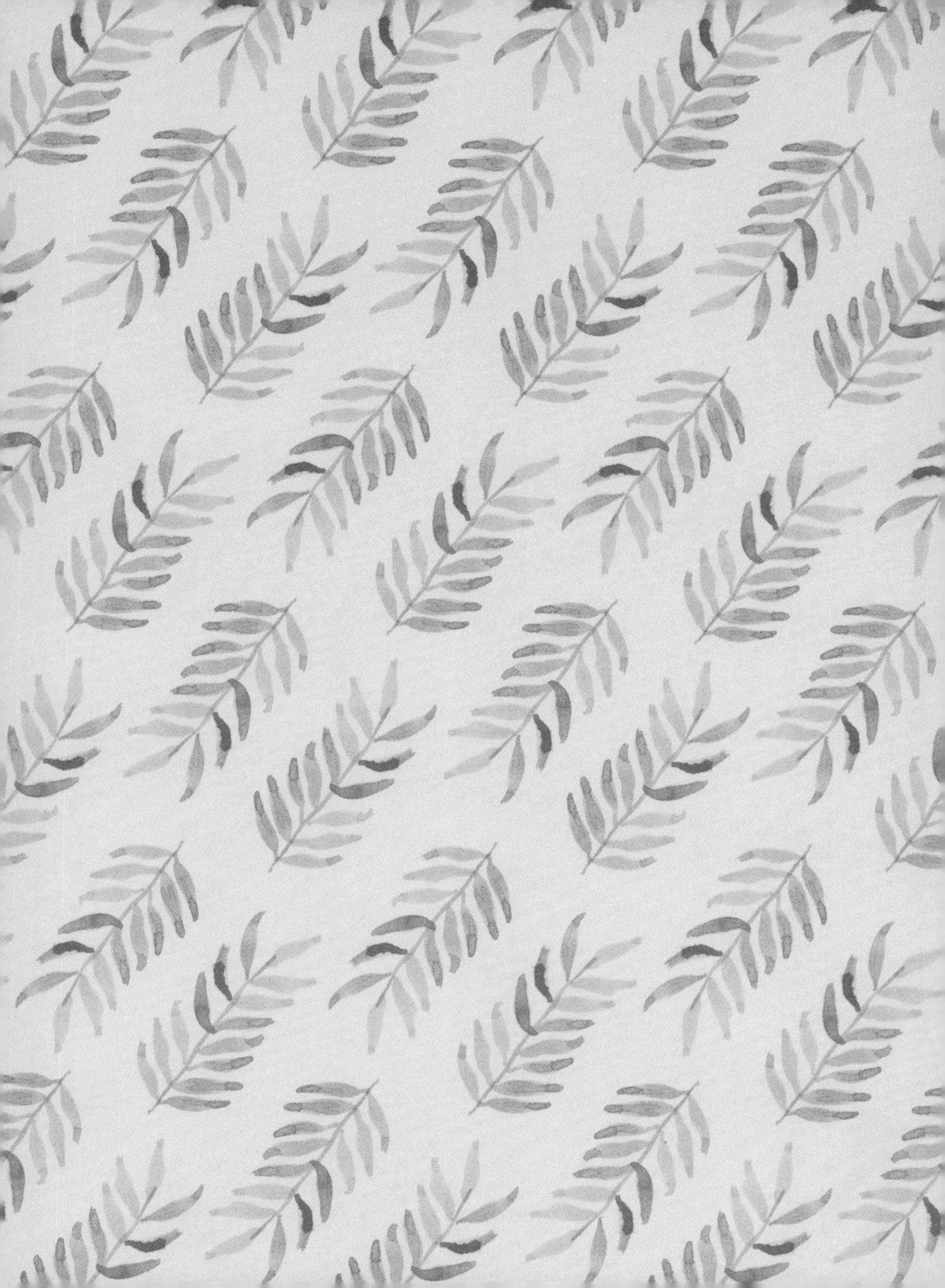

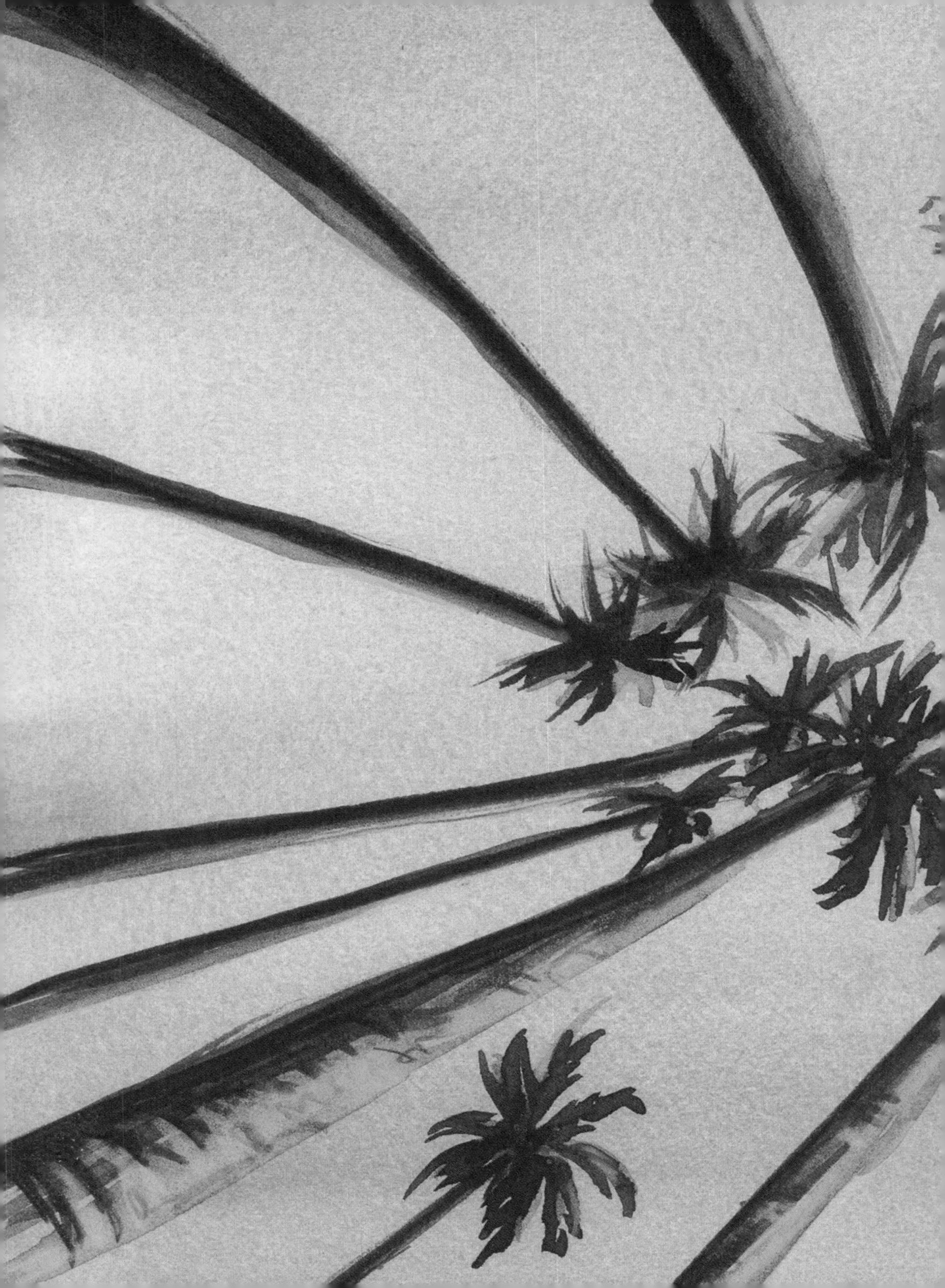

SEEKERS

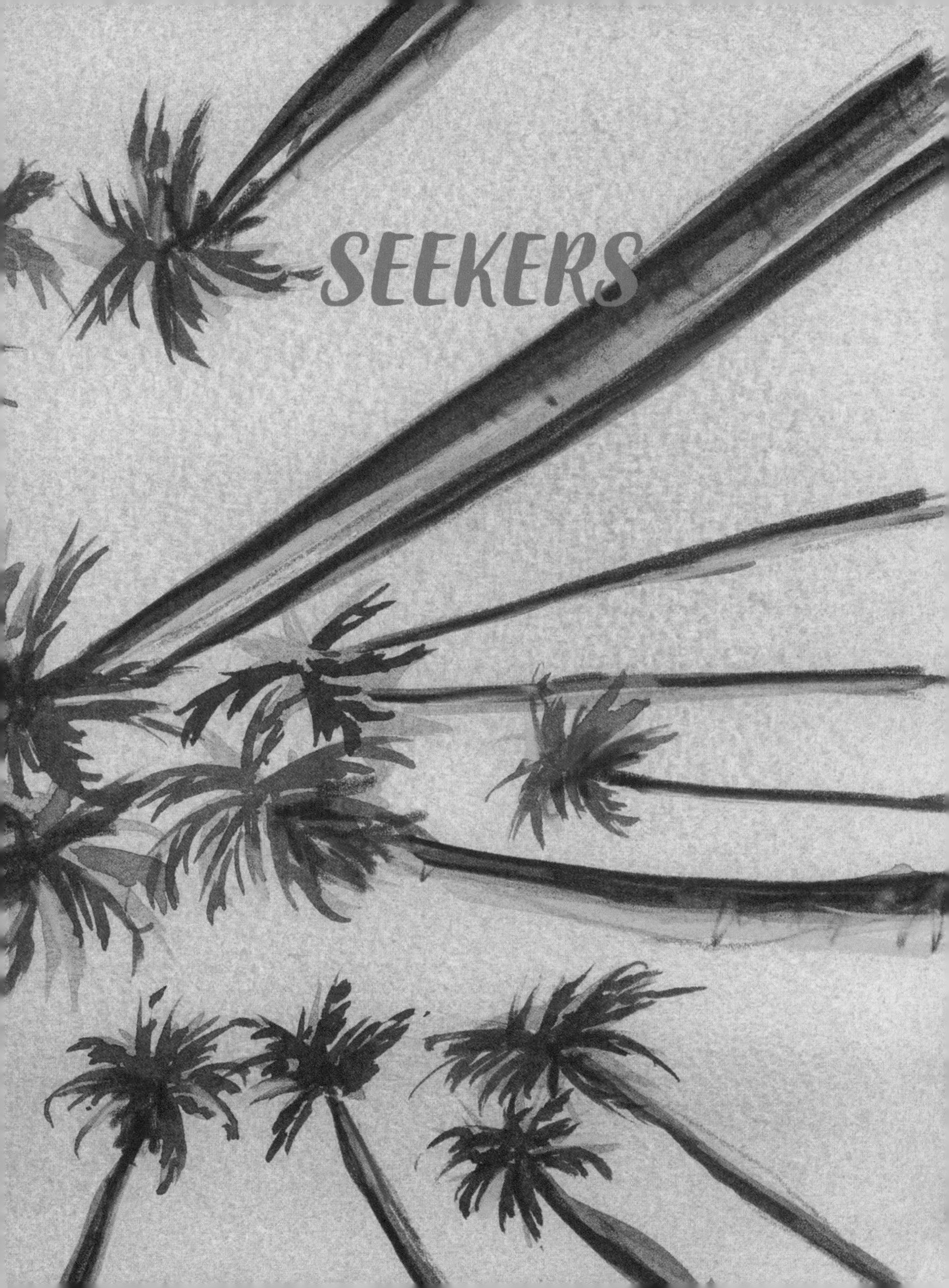

"EVERYONE TELLS ME OF ALL SORTS OF DELIGHTFUL POINTS TO WHICH I HAVE NOT YET GONE...IT IS TERRIBLY TRYING TO A GREEDY COLLECTOR LIKE MYSELF!"

THE EXPLORER
YNÉS ENRIQUETTA JULIETTA MEXÍA
Berkeley, California

IMAGINE FINDING YOUR LIFE'S PASSION AT THE AGE OF FIFTY, EM-barking upon a course of study at a university, and then traveling across remote and uncharted areas of the world, often alone, to pursue your work.

Now imagine it's 1921, and you're a Mexican American woman who'd just spent ten years in a sanatorium.

Ynés Enriquetta Julietta Mexía never collected trees in the sense of planting them on a property she owned, but she was a collector of specimens in the pursuit of science. In this sense, any botanist-explorer could be called a collector. But Mexía's work is remarkable for the scope, daring, and sheer volume of what she collected, not to mention the obstacles she overcame to accomplish it.

She was born in 1870 in Washington, D.C., where her father was a diplomat at the Mexican consulate. Her parents divorced, and she spent her childhood in the United States with her mother. But in 1896, she moved to Mexico to take care of her ailing father and assumed responsibility for his ranch after his death. Her time in Mexico was marked by tragedy: she struggled to manage the ranch, her first husband died, and her second marriage was so unhappy that she spent her days in her bedroom, curled in a ball, miserable.

In 1909, she moved to San Francisco to seek help for depression. Her letters to her husband stated quite plainly that she was unable to "bear with the marriage relation" and that she "had never known any sex love, nor do I think I am capable of it." What looked like depression might simply have been the effects of living under the expectations placed upon women at the turn of the twentieth century: forced into a relationship she couldn't tolerate, with no interests or occupation to make life worthwhile. She was unable to continue with her life in Mexico and permanently separated from her husband.

She moved to the Arequipa Sanatorium, a Bay Area women's facility run by Dr. Philip King Brown, who believed that women's lives could be improved by finding useful pursuits. It was under Brown's care that Mexía discovered an interest in botany that brought her back to life.

She became one of the early members of the Save the Redwoods League and the Sierra Club. In 1921, at the age of fifty-one, she studied botany at the University of California, Berkeley. A few years later, she returned to Mexico, this time as a plant collector on an expedition organized by Stanford University. She quickly realized that she preferred to work on her own, so she broke away from the expedition, made her way down the Pacific coast, and collected an astonishing thirty-five hundred specimens on that trip alone.

Over the next thirteen years she trekked throughout Latin America, often alone or with indigenous guides, riding horseback, sleeping outdoors for weeks or months at a time, and enduring every kind of calamity, from floods to earthquakes to a near-poisoning from deadly berries. On an expedition to Alaska, she was the first botanist to gather specimens in what is now Denali National Park. Her enthusiasm was insatiable: she once wrote, "Everyone tells me of all sorts of delightful points to which I have not yet gone . . . it is terribly trying to a greedy collector like myself!"

Botanist Nina Floy Bracelin, twenty years younger than Mexía, served as her assistant back home. While Mexía traveled, Bracelin cataloged her specimens, maintained her correspondence, and generally

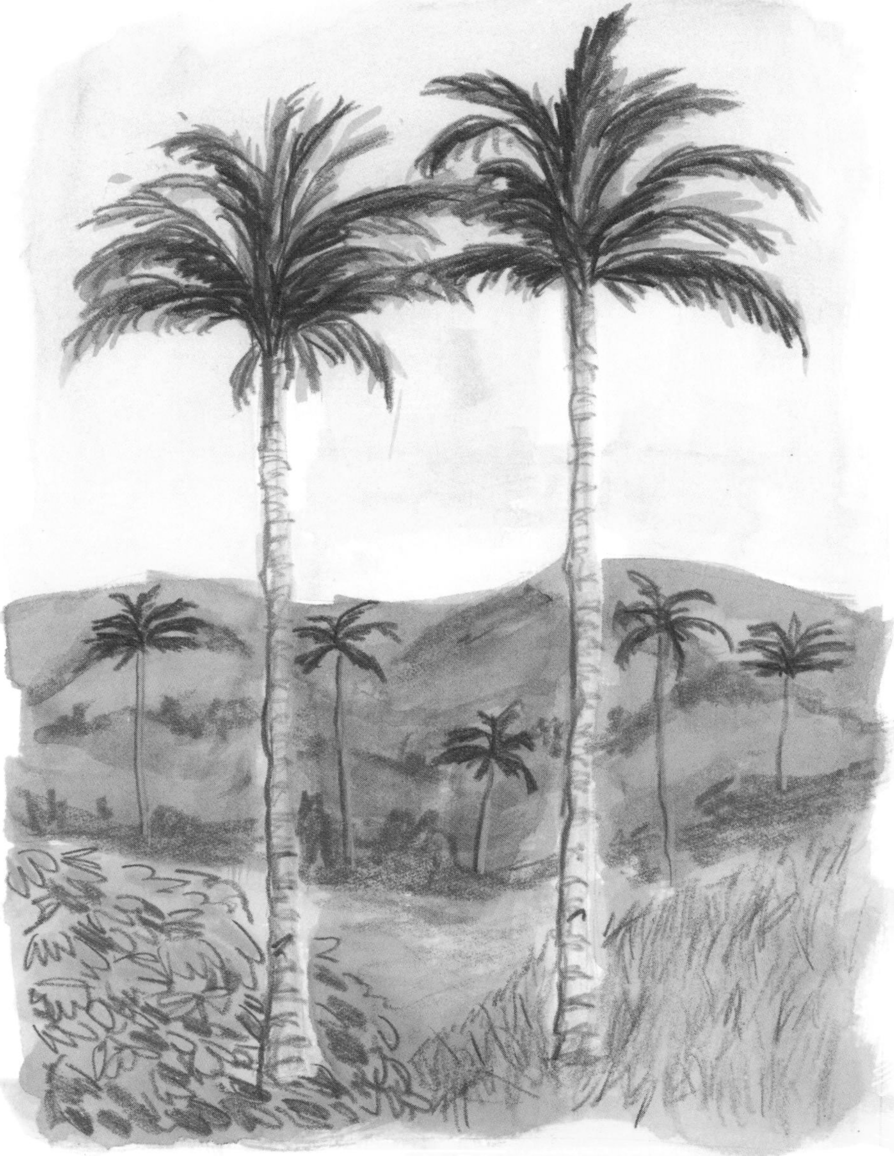

took care of the details so Mexía could explore. Their relationship was close and lifelong: Mexía made provisions for Bracelin in her will, and after Mexía's death from lung cancer at the age of sixty-eight, Bracelin continued to curate her specimens and preserve her legacy. There's nothing in the historical record to suggest that their relationship was romantic, however. Late in life, Bracelin told an interviewer, "She didn't care for people in the way that you and I do."

Although she began her botany career in midlife, Mexía collected a hundred fifty thousand plant specimens, including five hundred new species. Fifty plants are named after her, including the Ecuadorian and Colombian palm tree *Ynesa colenda*. Her writings, photographs, and specimens are held at major museums and libraries around the world.

On a 1937 expedition north of Quito, she sought the hard-to-find wax palm, *Ceroxylon quindiuense*. At the end of a particularly difficult day of navigating narrow trails on steep slopes, a gorgeous wax palm rose before her. She rushed over to take photographs, collect specimens, and make measurements and notes. "Then we started on the long journey back," she wrote, "arriving after dark, very tired, very hot, very dirty, but very happy."

PLANT EXPLORERS AND THE TREES NAMED FOR THEM

JOSEPH BANKS
(1743–1820)

British botanist who accompanied Captain Cook's voyages, among others. The Australian coast banksia, *Banksia integrifolia,* was collected by him in 1770.

DAVID DOUGLAS
(1799–1834)

Scottish botanist who traveled across North America, particularly in the Pacific Northwest, where he identified and named the Douglas fir, *Pseudotsuga menziesii.* The tree's common name belongs to Douglas, but the Latin name underwent several revisions before finally being named after another Scottish botanist, Archibald Menzies, who had described it three decades before Douglas did.

ALEXANDER VON HUMBOLDT
(1769–1859)

German scientist and explorer who traveled widely throughout the Americas. The Humboldt oak, *Quercus humboldtii,* was named after him by fellow explorer Aimé Bonpland, after they identified the tree together in the Andes.

MANUEL INCRA MAMANI
(D. 1871)

Bolivian plant hunter who worked alongside British explorer and rancher Charles Ledger. Mamani endured years of hardship, and was ultimately imprisoned and beaten to death, for assisting Ledger by locating sources of medicinal quinine bark. He was not credited for his sacrifice: the glory went to Ledger, for whom the quinine or Peruvian bark tree, *Cinchona ledgeriana,* was named.

FRANK MEYER
(1875–1918)

A Dutch-born American citizen who was sent to China by the USDA to find new plant species that might be of some benefit to farmers. The Meyer lemon, *Citrus × meyeri,* was one of his introductions.

MARIANNE NORTH
(1830–1890)

British artist and botanist who traveled widely and created bold oil paintings of the plants she saw. In the Indian Ocean archipelago of Seychelles, she collected samples of a flowering tree and sent them to Kew Gardens, where director Joseph Hooker named the tree *Northia seychellana* in her honor.

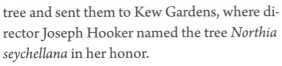

"MY ANSWERS WERE IN
THE TREES ALL ALONG."

THE GENE COLLECTOR
LUCAS DEXTER
Angwin, California

LUCAS DEXTER GREW UP AROUND HIS FATHER'S TREE COLLECtion. "My dad has a degree in plant biology," he said. "He started putting rare conifers and maples in the ground when I was about four. They grew into full-sized trees, and I grew into a full-sized person alongside them."

He didn't fully appreciate the botanical wonderland he'd grown up in until he returned home after a bruising Silicon Valley failure. He lost his job and the company he'd founded. "I was evicted and my car was impounded. My girlfriend left. The following year was my darkest. Denial, drinking, and ultimately depression set in. Eventually, I gave up the city, drinking, and searching in the wrong places. My answers were in the trees all along."

He returned to work in his father's landscaping business. This time, he understood what fascinated his father about trees, and what a tree could mean in a person's life. "When I bought my first home, my father gave me a seedling that he'd found under one of his Japanese maples," he said. "The original tree had a bit of variation in the leaf, but the offspring was almost entirely variegated and bright pink. He'd been growing it for ten years. So he gave it to me and named it Allis, after my mother. And then, that first summer, I killed it."

To lose such a heartfelt gift was devastating. He searched his father's property, hoping to find another seedling like it. Although he found plenty of interesting variegated Japanese maples that had sprung up over the years, there was never another like Allis. He regretted that he hadn't preserved the tree's genetics before he lost it.

He wasn't going to let that happen again. He learned to graft Japanese maples and, from there, began collecting unusual tree mutations he noticed as he went about his landscaping work. "I always have my eyes open, looking for mutations or variations, even as I'm driving down the road," he said. "I'll pull over if I see a variegated or golden leaf that stands out to me."

What Dexter looks for is a sport, or a genetic variation that only affects part of the tree. As a branch grows, cells inside the plant tissue might undergo a spontaneous mutation that causes the new growth to look slightly different. This is often how variegated leaves come about: the plant starts creating cells that don't produce chlorophyll, leaving white stripes and patterns in the leaves. (Nectarines are sports: they were discovered when peach trees produced a few mutated fruits with no fuzz on the skin. Even today, a peach tree will occasionally produce one or two nectarines.)

These mutations don't run through the entire plant. They tend to be limited to the growth that occurred after the mutation took place. So if Dexter sees an unusual color or pattern on a few leaves, he can only propagate that sport if there's enough new growth to clip.

Those cuttings have to be carefully rooted or grafted to another tree in order to survive. Some trees, such as oaks, are notoriously difficult to graft. In some cases, Dexter will send the cuttings to a specialty nursery that has the right equipment and expertise. If the propagation is successful, he might try to register the new cultivar through an international plant society, give it a name, and possibly sell it or offer it to other collectors to grow.

"I didn't realize when I started this that you could introduce something new to the world," he said. "When I came to work for my family's landscaping company, there was a fulfillment that tech just doesn't have, and that's building something that's permanent, and that actually grows, so it can increase in beauty and size. That fulfillment was wonderful, but what I lost was doing something really new and cutting-edge. The opportunity to play around with these genetics gives me that again."

He's introduced a dozen new plants to the world, including an unusual yellow oak, called *Quercus lobata* 'Dexter's Gold', from a sport he discovered in Napa Valley, high in a tree's canopy, and *Acer macrophyllum* 'Tubbs Fire', a variegated maple found after the devastating 2017 fire in Calistoga.

His tree collection, then, looks less like an arboretum and more like a laboratory, with all of his selections in pots, being fed by drip irrigation. The entire plot is only six feet wide and 150 feet long. "Someone else will grow them to maturity," he said. "I'm just after the genetics."

CHAMPION TREES: A SHORT HISTORY OF THE QUEST TO FIND TALL TREES

In the quest to find the world's largest trees, a new kind of tree collecting was born. Fortunately, today's big-tree hunters travel with cameras and measuring tape, rather than chainsaws.

1853 • Prospectors felled a giant sequoia in what is now the Calaveras Big Trees State Park in California. At the time, no one could believe that such massive trees existed. Photography wasn't yet sophisticated enough to do them justice, and the drawings of it seemed implausible. To prove it existed—and to make some money—a cross-section of the so-called Discovery Tree went on tour, like a circus sideshow.

1854 • Another sequoia was stripped of its bark (which later killed it), and the bark went on tour. This led to a public outcry against the destruction of giant trees. The modern conservation movement was born. But the fascination with massive trees had taken hold.

1867 • A report circulated in newspapers around the country about a geologist in Missouri, Professor G. C. Swallow, who had completed a geological survey of the state and recorded the diameters of the largest trees he found, including "a sycamore in Mississippi County, sixty-five feet high, which, two feet above the ground, measures forty-three feet in cir-

cumference." Not to be outdone, other state geologists started reporting on their largest trees, and cataloging champion trees became a popular civic endeavor.

1921 • The Pennsylvania Department of Forestry announced its Big Trees of Pennsylvania contest to find the largest of one hundred different species of trees in the state.

1932 • Iowa's Federated Garden Clubs held a statewide champion tree contest and awarded champion tree designations to an oak, an elm, and a maple, among others.

1940 • The nonprofit American Forests established the National Register of Big Trees to catalog every champion tree in the country. The largest tree by volume in the world, a giant sequoia (*Sequoiadendron giganteum*) in California known as General Sherman, with a trunk circumference of 102 feet, was immediately nominated to the registry and still holds the title today.

1998 • The tallest tree, a coast redwood (*Sequoia sempervirens*), was added to the registry with an impressive height of 321 feet.

2015 • The smallest champion tree in the registry, a southern bayberry (*Morella caroliniensis*), won with a height of fourteen feet and a crown spread of ten feet, which is small by comparison to other champion trees but still makes it a giant among southern bayberries.

2019 • The widest tree, a live oak (*Quercus virginiana*) in Georgia, was recorded with a crown spread of 161 feet.

TODAY'S CHAMPION TREES

American Forests still maintains the National Register of Champion Trees today. More than five hundred trees are listed, based on nominations from big-tree hunters across the country. Trees must be precisely measured by circumference, height, and crown spread, and then submitted to the state big-tree registry for confirmation. (State programs are often run by universities or state forestry departments.)

While the locations of most champion trees are kept secret to protect them, many state champion tree programs will advertise the locations of those trees in state parks and other public areas that are already well known.

Champion tree registries have also become popular in South Africa, New Zealand, Australia, Tasmania, and Canada, and across Europe and the United Kingdom.

HOW TO MEASURE A CHAMPION TREE, THE SIMPLIFIED VERSION

Big-tree hunters travel with laser range finders and other sophisticated instruments to precisely measure the size of champion trees. But here's a version you can undertake in your own backyard, using nothing but a measuring tape, a yardstick, a piece of chalk, and some stakes:

CIRCUMFERENCE • Measure the girth of the tree at four and a half feet above the ground.

HEIGHT • Mark the trunk four feet above ground with chalk (or tie a string around the tree). Walk away, holding up a yardstick, until that four-foot section measures one inch on your yardstick. From there, align the bottom of the yardstick with the base of the tree and measure where the top of the tree falls on the yard-stick. One inch equals four feet.

CROWN SPREAD • Stand under the tree and look up. Find the widest point in the crown and drop a stake. Walk across to the other side of the tree and drop another stake at the opposite end of the crown. Measure the distance from one end to the other. Do this again with the narrowest point in the crown. Take an average of the two distances to figure the average crown spread.

CALCULATE YOUR TREE'S CHAMPION TREE POINTS •
Circumference in inches + height in feet + average crown spread in feet = total points.

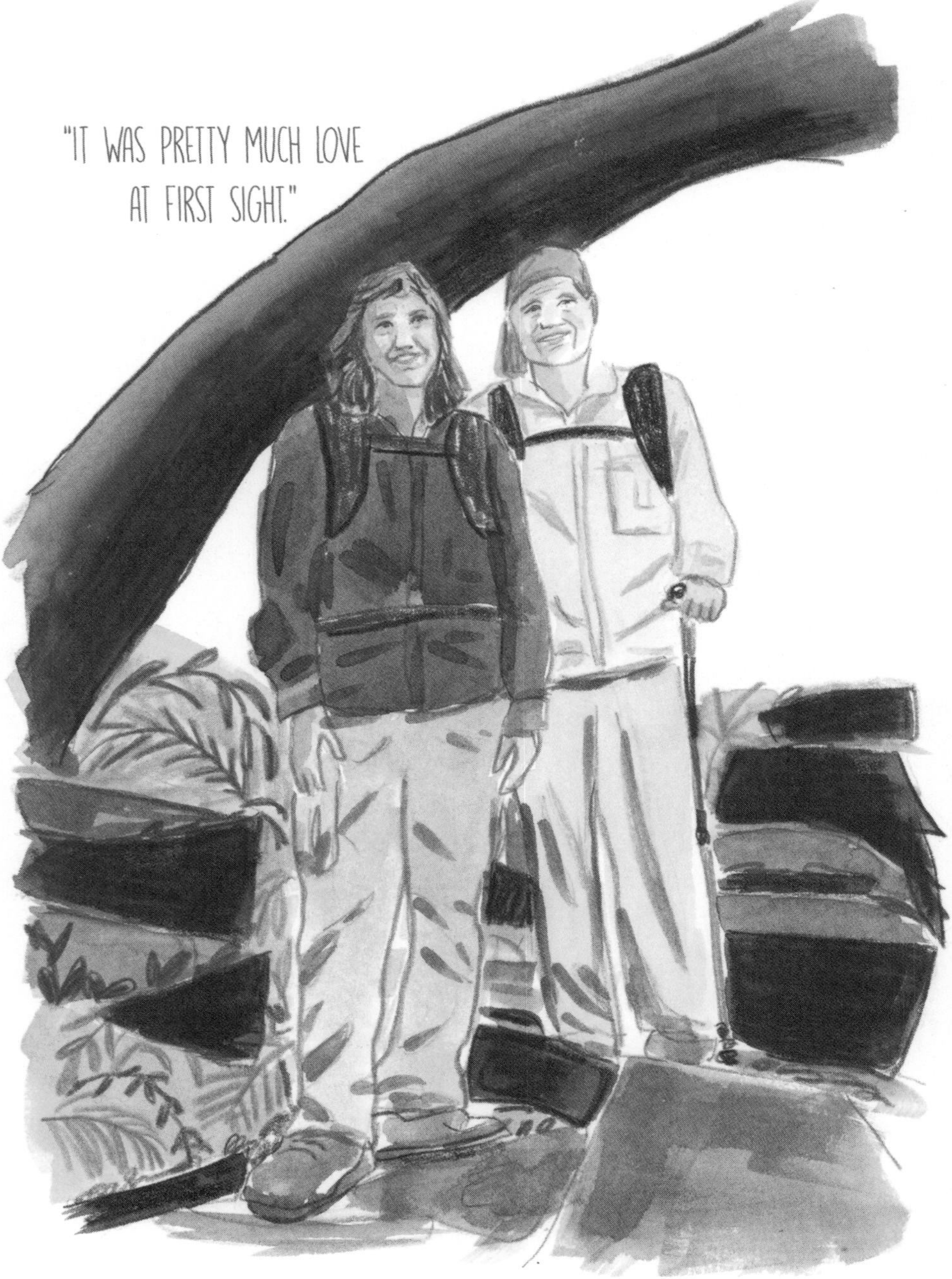
"IT WAS PRETTY MUCH LOVE
AT FIRST SIGHT."

THE CONSERVATORS
SUE MILLIKEN AND KELLY DOBSON
Port Townsend, Washington

TO HEAR KELLY DOBSON AND SUE MILLIKEN TELL THE STORY OF how they met is like watching the interviews with longtime married couples in *When Harry Met Sally*.

MILLIKEN: I studied botany and ecology in Vermont. I started a nursery up there.

DOBSON: And I was running a nursery out in Washington State. I would get these orders from this woman in Vermont who always bought the craziest, most obscure, wacky plants in my catalog, and I would think, *How are you going to grow these in Vermont?*

MILLIKEN: It turned out we had a mutual friend who was organizing a seed-collecting trip to China in 1997, and she invited us both to go.

DOBSON: I called another friend who had just been to China, to get his advice. He told me that if I had a seed-collecting partner in the field, someone to help take notes and so on, we could collect half again as much together as we could separately. And he told me that of everyone going on this trip, Sue would make the best collecting partner.

MILLIKEN: He likes to take credit for setting us up. But it was pretty much love at first sight. There were seven of us in the group, so the other five were watching this romance develop. We both felt like we'd known each other our whole lives.

DOBSON: We were so far away, outside of our normal routines. We weren't putting on airs, we were just being totally ourselves and walking around in awe of these amazing plants.

MILLIKEN: So we came back home, and upended our lives, and I moved across the country.

Together they started what is now the nonprofit Far Reaches Botanical Conservancy, with the goal of collecting and conserving rare and endangered trees and other plants. "Running a for-profit nursery—which is a loose term in the nursery business—wasn't lighting us up," Milliken said. "It was really the joy of saving and sharing these plants that we were after."

Being a nonprofit makes it easier to get permits to collect seed in other countries. It also allows Milliken and Dobson to work with other tree collectors, who want to see their rarest specimens grown in a few different places around the world as a kind of insurance policy. "This is especially important to those people in their eighties who just aren't sure what's going to happen to their collection after they're gone," Dobson said.

Their nursery spans just six acres in Port Townsend, Washington. "It's not an ideal site," said Milliken. "There wasn't a tree on the place when we bought it. It was very windy, and located in a frost pocket." They've had to build a lot of raised beds and special shelters to protect plants that aren't accustomed to cold Pacific winds.

While anyone can shop at their nursery, many of their best finds go to botanists doing research or trying to complete a collection. A walnut relative, *Rhoiptelea chiliantha,* is an early ancestor to modern pecans and hickories. It was eagerly snapped up by a botanist working for the USDA who planned to do genome sequencing that he hoped would have implications for the future of pecan tree breeding.

A tough, sinewy tree that grows at some of the highest elevations in the world, *Polylepis lanata* had hardly ever been cultivated outside of its native range in Bolivia. "We heard from a nursery in California that was selling seed shares to finance an expedition in South America," Dobson said. "A lot of times you buy seed shares and you just get whatever they collect. But we sent them a couple hundred bucks and said, '*Polylepis.* Nothing else.' And they got some!"

Those are the prized trees in their collection now. "They can get to eighty feet in the wild, but they're really heavy and tortuous," he continued. "They grow in an extreme environment, so they have this thick papery bark that helps protect them. It's an amazing tree that more people should know about."

Another extraordinarily rare tropical tree native to China and Vietnam, *Diplopanax stachyanthus,* was only known in the United States from fossil records. Although it had been described in the botanical literature, it was so poorly understood that taxonomists constantly moved

it from one plant family to another. "We think we had the first specimen to flower in cultivation," Dobson said. "Of course we had a party when it bloomed. Those are the kinds of parties we have."

While it's a thrill for plant geeks to grow these extraordinarily rare trees, the broader purpose of the conservancy is to safeguard species under threat from human activity and climate change. If they can't be protected in their native country, the conservancy will make a home for specimens where they can be propagated for a network of international researchers. "It makes us feel like we're contributing to the world," Dobson said, "and making some very small atonement for the idiocy of the human race."

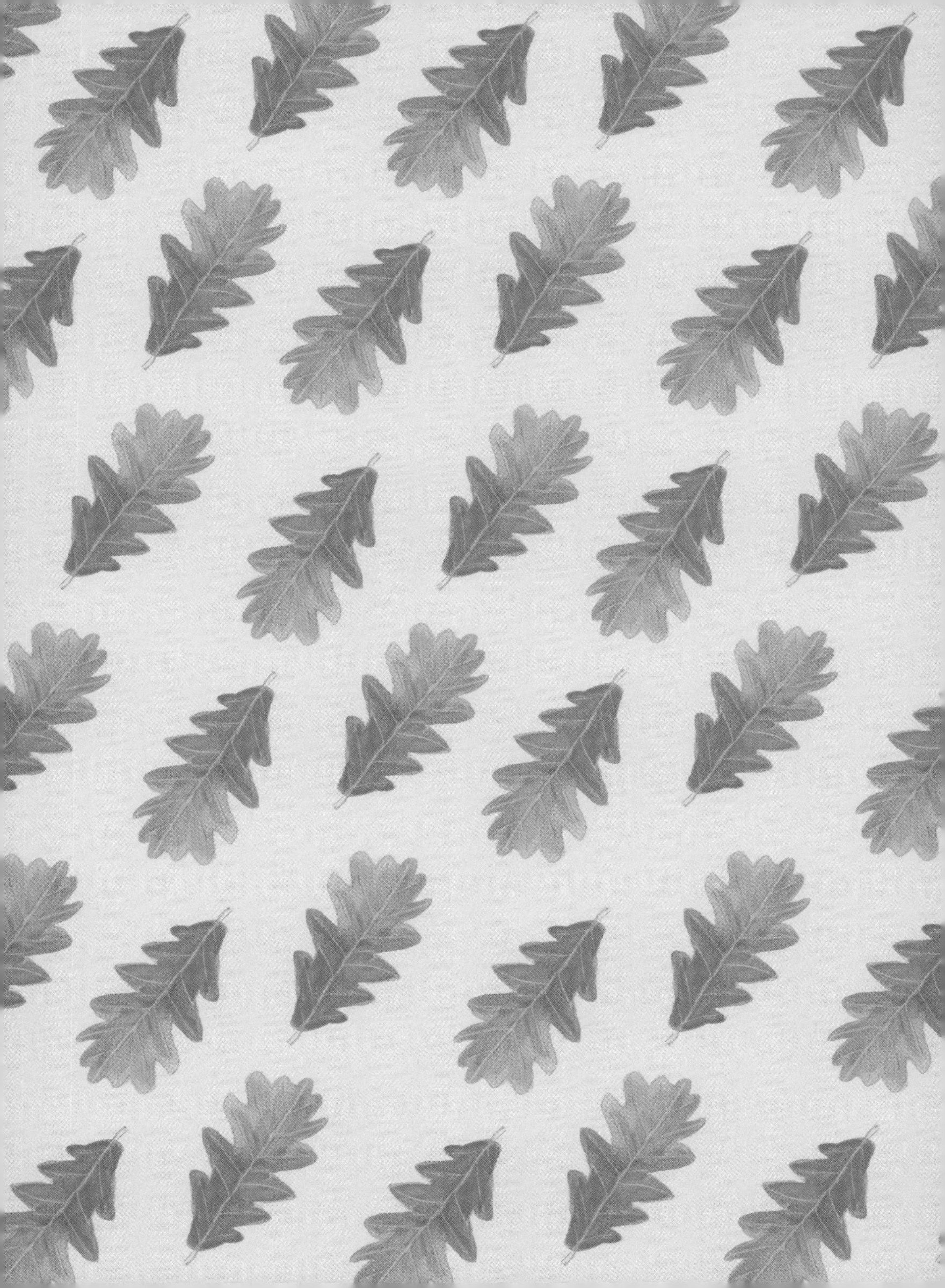

"THIS TREE HAS
GROWN FROM SEED
THAT'S TRAVELED
FARTHER THAN MOST
OF US EVER WILL."

THE ASTRONAUT
STUART ROOSA
Arlington, Virginia

As a smoke jumper for the U.S. Forest Service in 1953, Stuart Roosa was in the business of saving trees. He was stationed in southern Oregon with a couple dozen other jumpers. If a fire was spotted, he and his fellow crew members would be flown overhead and dropped from the plane via parachute. The trees themselves posed a hazard: accidentally landing atop a two-hundred-foot-tall Douglas fir was almost as dangerous as dropping into the middle of a fire. Once they landed, their mission was to put the fire out with little more than shovels and dirt. Roosa's crews tended to land safely and get the job done. After the fire was extinguished, they'd hike to the nearest road, equipped with a map, a compass, a canteen of water, and a little food.

A mission like this requires guts, careful planning, and survival skills, all of which served him well when he was accepted into NASA's astronaut program just over a decade later. He'd received his pilot training in the air force, but over the years he'd kept in touch with friends from his smoke-jumping days. When the news broke that Roosa would be on board the 1971 Apollo 14 mission, Forest Service officials reached out and asked if he'd consider taking some tree seeds to the moon.

Astronauts are allowed to carry a personal preference kit filled with whatever small tokens would serve as a good souvenir of their time in space: flags, family photos, commemorative stamps, and the like. Roosa carried canisters holding sealed bags of seeds from five trees: American sycamore, loblolly pine, sweetgum, coast redwood, and Douglas fir. Be-

244 • Amy Stewart

cause Roosa stayed on board the command module, the seeds didn't actually touch the surface of the moon. But they went into space, affording the Forest Service the opportunity to conduct a science experiment that would also win them some publicity and goodwill.

After the spacecraft splashed down in the Pacific Ocean, the astronauts were quarantined and their possessions put into a vacuum chamber for decontamination. Unfortunately, Roosa's seed bags had been removed from their canisters, and the bags exploded during this process. NASA historian John Uri said, "We've looked for photos of this incident, but I don't think any were taken. Now we document everything. But in those days we had this finite resource called film."

The seeds were hastily gathered up and sent to a USDA scientist working for NASA, who couldn't get any to grow in Houston. He sent the remaining seeds on to Forest Service greenhouses in California and Mississippi, where they were able to germinate the seeds, along with a control group kept on Earth for comparison purposes. (There were no differences between the trees sprouted from moon seeds and the ones from seeds that stayed behind.)

What happened next to those moon trees remains a bit of a mystery. The Forest Service kept some records that suggest saplings were sent to each state for the nation's bicentennial. Newspaper clippings from 1976 documented some ceremonial tree plantings: a redwood was planted near the entrance to California's capitol building in Sacramento, and a sycamore went into Philadelphia's Washington Square. Universities, museums, and, of course, NASA facilities received trees. Some were marked with a plaque, others were not. Some lived and some died.

After the bicentennial, the moon tree story faded away. Even NASA largely forgot about it. Then, in 1996, just two years after Roosa died, NASA planetary scientist David Williams got a call from a third-grade teacher in Indiana. "Her class was doing a project on historic trees," Williams said, "and there was this tree at a Girl Scout camp with a sign about one of the Apollo missions. So I asked around here at Goddard and talked to some of the old-timers, but nobody remembered it. Our history office had a folder with a little information. I thought it was a cool story. This was back in the early days of the web. So I made a page about the moon trees."

He was struck at once by what a clever idea it had been to carry seeds into space. "You know, you could fit every single thing they brought back from the moon in a few boxes in your garage. Except for these trees. He took these little seeds, but that was something that would grow. Now they're bigger than anything you could carry to the moon and back."

Once he put a little information out there, people started contacting him with their own moon tree sightings, which he added to the website. Then the Forest Service contributed some records from their archives, which helped to trace the fate of these trees.

Now he keeps a list of second-generation moon trees, germinated from the descendants of the originals. Some have been propagated by university botany departments, and historic tree nurseries have offered them for sale. "I have a little second-generation moon tree in my backyard," Williams said. "There's hundreds of them out there. The notion that this tree has grown from seed that's traveled farther than most of us ever will, there's just something wonderful about that."

MOON TREES YOU CAN VISIT

Moon trees come and go. Previously unrecorded trees might still be discovered, and well-known trees fall victim to storms, disease, and old age. Some are on private property or otherwise difficult to find. Here are just a few that are planted in public spaces, clearly marked, and eager to receive visitors.

U.S. FOREST SERVICE OFFICE, TELL CITY, INDIANA • The sweetgums don't seem to have fared as well as the other species. A pair of sweetgum moon trees grow in front of the garage at this Forest Service office and are easily visible from the road, but those are the only ones known to have survived outside of private collections.

BIRMINGHAM BOTANICAL GARDENS, BIRMINGHAM, ALABAMA • A sycamore thrives next to the rose garden. Just ask the staff, and they will point the way.

FRIENDLY PLAZA, MONTEREY, CALIFORNIA •
A coast redwood stands prominently at the
southern end of the plaza and is marked
with a plaque.

**WASHINGTON STATE CAPITOL, OLYMPIA,
WASHINGTON** • A Douglas fir was planted
on the capitol grounds, near the Tivoli
fountain, in 1976 and rededicated with a
plaque in 2003.

**HISTORIC WASHINGTON STATE PARK,
WASHINGTON, ARKANSAS** • A loblolly pine is
planted near the historic courthouse and
marked with a plaque that designates it as an
Arkansas Famous and Historic Tree.

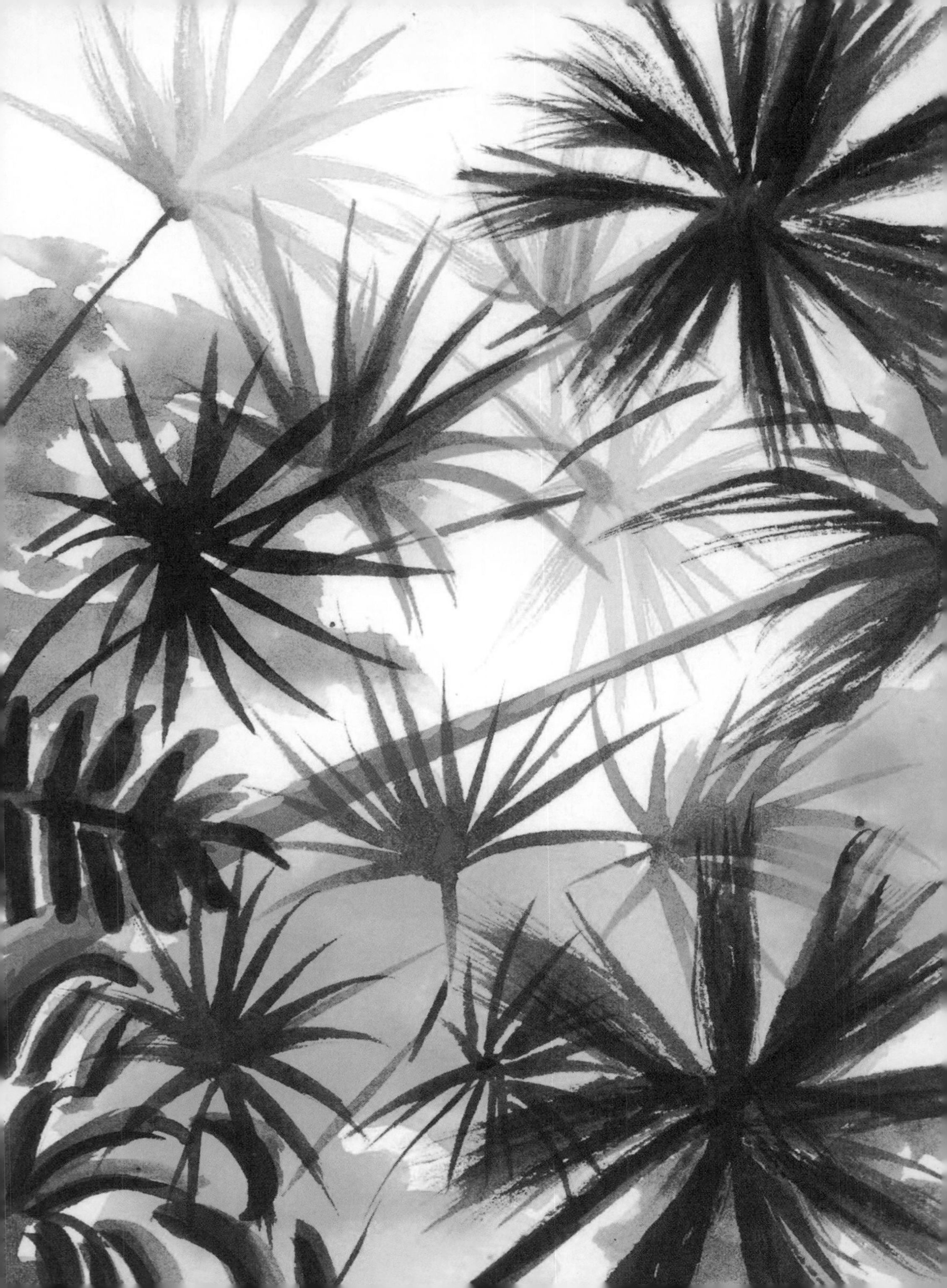

PRESERVATIONISTS

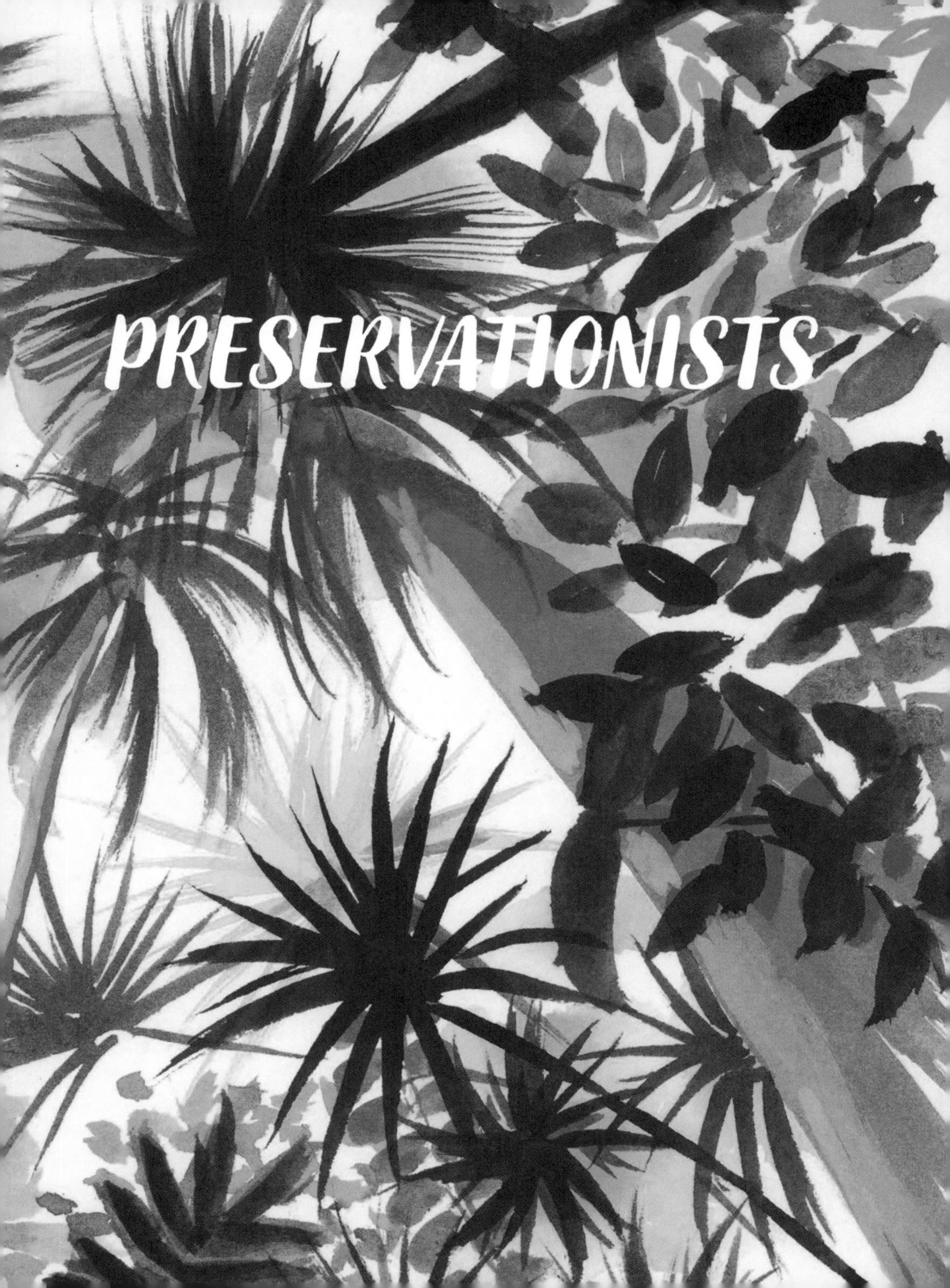

"I'M RELATED TO CHESTNUTS
THROUGH MY MOTHER."

THE CHESTNUT CHAMPION
ALLEN NICHOLS
Laurens, New York

ALLEN NICHOLS STUDIED AGRONOMY AND BIOLOGY IN COLLEGE, but when he took a job as a lineman for a local utility company, he found that the work suited him. "It's nice to be outside and to hear those geese when they're flying," he said. "You don't realize what you're missing when you're not working outside."

He and his wife live on sixty acres of woodland in Laurens, New York. He's planted fruit and nut trees, started oaks from acorns with his young daughters, and managed a woodlot for firewood. But all the while, he had his eyes on the chestnut trees.

It's hard to overstate the importance of the American chestnut (*Castanea dentata*) to eastern forests. Roughly four billion chestnuts once dominated the landscape. A staggering array of insects depended upon the trees, including several dozen species of moth alone. Their nuts fed birds, squirrels, and bears. A farmer could run hogs and cattle through the forest to let them forage for their dinner and still collect enough chestnuts for a winter feast. The trees were massive, with trunks spreading over ten feet in diameter, and they lived for five hundred years. Their wood was dense, straight grained, and useful for everything from furniture to fence posts.

The trees' reign ended in 1904, when a batch of imported Japanese chestnut saplings came ashore, along with a disease to which they were naturally resistant. The American trees had no such immunity and were wiped out within a few decades. American chestnuts still sprout from old stumps today, but the blight quickly finds them.

Nichols's parents felt this loss keenly. "My mother grew up on a farm that had a lot of chestnuts on it. They all died. My father would go out with his uncle and cut fallen chestnuts for fence posts after the blight got them." Three generations of his family mourned the death of the trees.

Maybe that's why he was so drawn to the work of the American Chestnut Foundation. The group was founded in 1983 to try to address the loss of this majestic tree. By the time Nichols got involved, around 1990, they were well on their way to breeding a cross with a resistant Chinese species in order to continue breeding that cross back into the American trees until the new hybrid retained only the disease resistance of its Chinese ancestor.

But this has proven tricky. The genetics are complicated. That's why scientists at SUNY's College of Environmental Science and Forestry began working on another approach. They isolated a gene found in many other plants that can break down oxalic acid, which is the very substance that the chestnut blight uses to destroy trees. By taking that gene from an unrelated plant—in this case, wheat—and inserting it into the American chestnut in the laboratory, they could begin to breed a resistant tree.

This project, which was taking place only a couple hours' drive from his house, appealed to Nichols. In 2000 he volunteered his land for SUNY's research. Hosting these experiments is no easy feat: because transgenic breeding methods are involved, Allen's orchard requires a special government permit. He has to find ways to keep original American chestnuts alive in spite of the blight, which means spraying with fungicide. And he has to take extreme measures to make sure no pollen escapes, so that these experimental genetics aren't released into the wild.

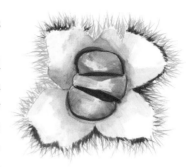

Every summer, when his two hundred chestnut trees begin to bloom, he climbs a ladder and puts bags over the flowers so they don't receive any local pollen. When the flowers open, he goes back and hand-pollinates each blossom with pollen from the transgenic trees kept under strict controls on SUNY's campus. In August, he places a wire enclosure over each pollination bag to protect the young chestnuts from predators. When the nuts are ripe, he cuts off the branches just below the bags and returns them, intact and numbered, to SUNY's laboratory.

The idea is to keep outcrossing the transgenic trees with the old American chestnuts. "It's like looking at a family tree," he said. "You don't want everybody coming from the same father. You want generations, and cousins, and lots of diverse offspring."

If the transgenic trees reproduced in the wild, not every offspring would be blight resistant. Some wouldn't inherit the gene from their parents. "There would still be quite a bit of diversity," he said. "A nursery could sell a tree that's guaranteed to be blight resistant, but when it pollinates a wild tree, only half of the nuts will get the blight-resistant gene."

All of this is hypothetical until the transgenic trees pass a strict regulatory review. If they do become available, Nichols has plans for them. "I'd love to plant them along my mother's old property, right by the road, so anybody could collect the nuts and enjoy them," he said. "She loved those trees. I'm related to chestnuts through my mother."

"DOESN'T EVERYBODY HAVE A
HUNDRED CAMELLIAS?"

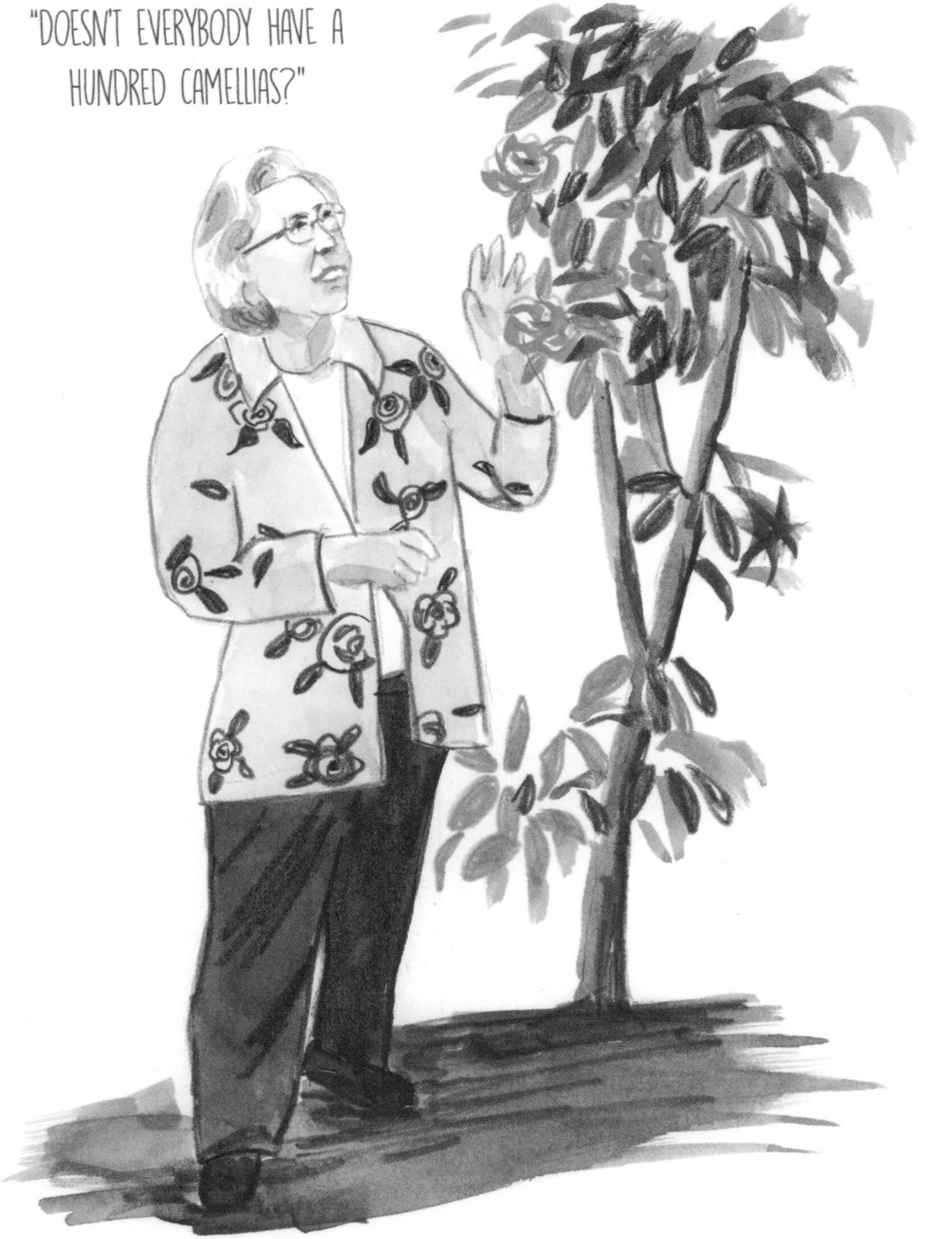

THE CAMELLIA PRESERVATIONIST

FLORENCE CROWDER

Denham Springs, Louisiana

FLORENCE CROWDER GREW UP AROUND HER FATHER'S COLLEC-
tion of camellias, but she never paid much attention to them. This was
just after World War II, when her parents built a new home in Denham
Springs and started planting camellias and azaleas. "They would drag us
kids to these parks and nurseries and gardens on weekends," she said,
"and they ended up planting a hundred camellias on their property. It
didn't really mean a whole lot to me. You know how kids are—so what?
Doesn't everybody have a hundred camellias?"

Her father was a carpenter who built homes all over town. When a
job was finished, he'd often give the new homeowners a camellia.
Sometimes, they would give him one. When her father died in 2005,
Crowder and her sisters inherited a house planted with camellias that
were now over fifty years old. For the first time, she got curious about
them.

"I wanted to learn the names of them, and to post the names on the
plants. One thing led to the other, and as I needed information, what
did I do but join the Baton Rouge Camellia Society? That's what got me
hooked."

She started attending camellia shows and meeting other collectors.
She also dug into the American Camellia Society's registry, which dates
back to 1948 and includes more than three thousand cultivars. Then
she discovered the worldwide registry of the International Camellia
Society—and that's where it got interesting. Their list documents ca-

mellias dating back to the nineteenth century, though a few Japanese and Chinese cultivars were recorded as early as 1600.

Now she had a quest. She didn't just want to identify the trees in her father's collection. She wanted to find all the lost camellias from the nineteenth century: those that might never have been photographed, or whose existence was only recorded in a few lines of description on a registration form. About 450 of these had been registered in the United States, and that's where she decided to focus her efforts.

She didn't have much time left. Although a few noteworthy camellia specimens have lived for hundreds of years, most survive only a century or two. Any nineteenth-century trees were likely to be near the end of their life span.

Her search for lost camellias has taken her to England and across Europe, where like-minded collectors are also preserving the old cultivars. A network of camellia enthusiasts closer to home are also on the lookout. "Sometimes people have what I'm looking for, but they don't know it and I don't know it. All we have is a description, written in technical plant nomenclature, and those can be difficult to decipher. There could be twenty-five camellias that fit one description." If she can visit the tree, she'll take photographs and possibly cuttings, so that it can be properly identified and propagated.

She's managed to collect two hundred, most of which live in pots so that she can move them around and protect them from the summer sun. Although she's looking primarily for those registered in the United States, she'll take any nineteenth-century camellia she can get her hands on, from anywhere in the world. In 2017, she and her husband donated one hundred trees to Louisiana State University's Burden Museum and Gardens in Baton Rouge, which already had one of the largest collections of camellias in the United States. Now their Florence and Charles

Crowder Camellia Collection extends the reach of the university's holdings back into history.

Crowder, who serves as an unofficial historian for her hometown and manages the history room at City Hall, has no trouble explaining why these old cultivars are worth finding and preserving.

"Why?" she said. "Because they once were. If they were important enough to register, they should be important enough for us to keep them going."

THREE CAMELLIAS NAMED AFTER FLORENCE CROWDER

'Florence's Debutante'

'Florence's Fancy Formal'

'Florence Crowder'

SAVING JAPAN'S CHERRY TREES

The ethereal pink blossoms of flowering cherry trees are a harbinger of spring. For more than a thousand years, Japanese horticulturists bred the trees to produce showy flowers, with single or double petals, in a range of colors from white to yellow to deep pink. While the flowering varieties (consisting of many species in the genus *Prunus*) are related to edible cherries, they produce small, bitter fruit palatable only to wild animals.

Although the trees have been a source of pride in Japan for generations, they were in peril a century ago. Japanese municipal planting schemes favored just one variety, 'Somei-yoshino'. It was cloned and propagated throughout the country, and in the twentieth century, it became a symbol of nationalism and conformity. With few efforts to conserve the other three hundred varieties, they started to disappear. It took a handful of dedicated collectors to bring them back, and to encourage the Japanese public to celebrate diversity over uniformity in these beloved trees.

SEISAKU FUNATSU
CHERRY TREE EXPERT
1858–1929

Funatsu worked tirelessly to preserve the varieties that grew around the Arakawa River in Tokyo. He documented each one and commissioned artist Kōkichi Tsunoi to illustrate them so that they could be properly identified. Funatsu was so well respected as a cherry tree expert that he was asked to select the varieties that were given to the United States to plant along the Potomac River in 1912. He also donated illustrations by Tsunoi to accompany the gift; those works of art are now housed at the Smithsonian.

COLLINGWOOD INGRAM
CHERRY TREE COLLECTOR AND AUTHOR
1880–1981

Through trips to Japan and contacts with nurseries, Ingram planted a diverse collection of cherry trees in his garden in Kent, England. He met Funatsu in 1926, and both men were astonished to learn that Ingram had varieties growing in his garden that could no longer be found in Japan. This began a decades-long campaign on Ingram's part to restore those varieties to Japan, preserve them in England, and educate the public. He became a world-recognized authority on cherry trees.

DAVID FAIRCHILD

PLANT EXPLORER AND USDA RESEARCHER
1869–1954

In 1906, Fairchild imported an assortment of cherry tree varieties from Japan for a test plot in Maryland. At the time he wrote, "There were once three hundred varieties, but of these three hundred, many are almost indistinguishable. The striking kinds would probably not number more than thirty or forty." He encouraged Americans to plant the trees for their outstanding beauty in early spring.

ELIZA SCIDMORE

JOURNALIST, PHOTOGRAPHER, GEOGRAPHER
1856–1928

A bold and adventurous journalist who traveled extensively in Japan, Scidmore worked for twenty-five years to convince officials in Washington, D.C., to plant cherry trees along the Potomac. David Fairchild supported her in these efforts. She finally found a champion in First Lady Helen Taft. The first shipment of trees arrived from Japan in 1910, but they were infested with pests and had to be burned. Finally, two thousand healthy trees arrived in 1912, and Scidmore, at the age of fifty-five, was there to see them planted.

MASATOSHI ASARI

CHERRY TREE BREEDER
1931–

Asari began breeding new cherry varieties in Hokkaido, in northern Japan, in the 1950s. One hundred of his 'Matsumae' varieties are on display at a park of the same name in Hokkaido. Matsumae Park attracts visitors from around the world who come each spring to see the remarkably diverse collection of cherry trees in bloom. Because winters are colder there, these trees are also suited to the climate in Europe and the United Kingdom. Asari, who remembered the suffering of British prisoners of war in Hokkaido during World War II, donated a selection of his trees to Windsor Great Park in 1992 as a gesture of reconciliation.

"WE'VE SAVED ALL OF THESE APPLES. THAT MAKES ME HAPPY."

THE APPLE PRESERVATIONIST

JOANIE COOPER

Molalla, Oregon

JOANIE COOPER WAS AT A MEETING OF THE HOME ORCHARD SOciety, just south of Portland, Oregon, when a man named Nick Botner approached her. "He had developed cancer, and he was worried that his orchard would be lost," she said. "So he said, 'Joanie, you need to come out and buy my farm from me.' Well, I couldn't do that! But that's where it started."

Botner knew that Cooper was interested in old apples. She lived on a property that held a small orchard, and she'd joined the society to learn more about forgotten apple varieties. But she was in no position to buy his 125-acre farm, which included not just the apple orchard, but several hundred other fruit trees, a vineyard, grazing land for sheep, and a large home. "It was too much," she said. "But I knew we had to save that apple collection."

Botner grew an astonishing forty-five hundred different apple varieties, making his possibly the largest apple collection in the world. It was entirely likely that his collection held varieties that didn't exist anywhere else, and because an apple seed doesn't produce offspring identical to its parents, these old varieties must be grafted to be preserved. As Botner's health failed, he couldn't tend his orchard anymore. If his property was sold, the new owner might not be interested in the apples at all.

In 2011, Cooper and two partners formed a nonprofit called the Temperate Orchard Conservancy, with the aim of cloning and preserving

264 • Amy Stewart

the Botner collection. She found a piece of property near his farm that would be suitable for an orchard, and the painstaking work of identifying and cloning thousands of apple trees began.

Apple identification is never easy. Even if Botner's trees were labeled, the newly formed conservancy wanted to confirm the name and heritage of each variety as they took cuttings and planted them into their own orchard.

Apple collectors depend upon old nursery catalogs and farm bulletins to piece together names and identifying features of forgotten varieties. The USDA's pomological watercolor project, which employed dozens of artists to paint full-color images of fruit grown in the late nineteenth and early twentieth centuries, is an extraordinary resource. Some collectors pore over old newspapers to read the lists of winners at county fairs, hoping to find names and descriptions of old varieties.

Cooper depends on references like these to definitively identify the apples in Botner's collection, but she also looks forward to the day when DNA analysis can give faster and more decisive answers. Washington State University has launched an apple genome project that has sequenced more than three thousand varieties so far.

Botner died in 2020, and his family has so far held on to the farm. Meanwhile, volunteers with the Temperate Orchard Conservancy have managed to graft every apple tree in his orchard. There's still a great deal of work to do: more than a thousand young trees are sitting in pots until they're large enough to plant in the ground. Fields need to be prepared and fenced to accommodate those trees. The ongoing chores of an orchardist—irrigation, pruning, weeding, pest and disease control, and harvesting—will only grow as the orchard does.

Because of their work on Botner's orchard, Cooper and her partners have gained a reputation as apple identification experts. Other aficionados of lost and forgotten apples send fruit to them for identification. Because apples can only be identified when perfectly ripe, they're inundated with boxes of mystery fruit in the fall. "We're pretty good," she said. "About five percent of the time, we're not sure. But we usually get it right." Identification fees bring in a little money to support the orchard, and every winter they publish a list of scion wood (little branches that can be grafted) available for purchase. Otherwise, they rely on donations.

"It's a lot of work and there's never enough money," Cooper said, "but we've saved all of these apples. That makes me happy."

"I WASTED THIRTY YEARS OF MY LIFE TRYING TO BECOME FLUENT IN FRENCH. I WISH I'D BEEN ON THIS TREE PATH INSTEAD."

THE HISTORIC TREE COLLECTOR
VICKI TURNER
Nashville, Tennessee

VICKI TURNER WAS IN THE MIDDLE OF A BUSY CAREER AS A WINE broker when her mother decided to place a conservation easement on a thirty-three-acre family property north of Nashville. "That conservation easement just hit a button I didn't know was there," Turner said. "Once I knew it was never going to be a subdivision, I started to look at that land differently. I thought, *What better place to plant a tree sanctuary?*"

She began to study trees, with an emphasis on rare and endangered species. "I knew the trees I planted would never be mowed down by a bulldozer, which made it worthwhile to try to track them down. And that's not easy—they're rare and endangered for a good reason!"

Among her prized trees are a Georgia plume, *Elliottia racemosa,* a small flowering tree that grows wild in only a few spots in coastal Georgia. The corkwood tree, *Leitneria floridana,* is another treasure that wasn't easy to come by. It's more of a dense shrub than a tree and can be found in a few wetlands in southeastern Georgia. These trees aren't widely distributed in nurseries, but sometimes a few are sold to collectors as a way of keeping them alive outside of their threatened habitat.

Then she picked up a book on historic trees. "These are all descendants of famous trees, and trees associated with famous people," she said. "I knew right away that I wanted to do historic trees too. I wanted every one I could get my hands on."

The book, *America's Famous and Historic Trees,* was written by Jeffrey

Meyer, a nurseryman who ran a historic tree nursery for the nonprofit American Forests. By working with historic home foundations, he was able to collect seeds and grow trees that had some historical significance: a tulip poplar planted by George Washington at Mount Vernon, or a pin oak from Elvis Presley's Graceland. Although the American Forests nursery has closed, the concept has been taken up by a retired Tennessee couple, Tom and Phyllis Hunter, who offer a few dozen historic trees through their nursery, American Heritage Trees.

"I found out they were only forty minutes away, and I called them up and told them I had to come right over," Turner said. "I bought every different tree they grew. I just love the idea that these trees are being preserved as part of our history. I have a standing order for every new tree that comes in. I want them all."

She planted more than four hundred trees at her family's property outside of town, and then she turned her attention to the homeowners' association where she lives. With fifteen acres of common land, she was able to expand her collection. In addition to cultivating rare, endangered, and historic trees, the community often plants a tree as a memorial when a resident dies. Now both Turner's conservancy and the land she lives on are designated as certified arboreta through Tennessee's Urban Forestry Council.

"I take such delight in tree identification and in weirdo trees," Turner said. "As a wine broker, we were importing wine from around the world. I wasted thirty years of my life trying to become fluent in French. I wish I'd been on this tree path instead. It gives me so much more pleasure. I make up for lost time with the intensity of my study and my pursuit."

COLLECTING HISTORIC TREES

Tom and Phyllis Hunter, owners of American Heritage Trees, would be the first to say that it isn't easy to grow and sell historic trees. They've had to build partnerships with historic homes and other sites of historical significance. The relevant trees have to be identified, viable seed has to be collected, and then it takes years of work in a greenhouse to grow the trees to a size that allows them to be shipped and sold. The availability of trees can shift from year to year. "We're lucky if we can add one or two new trees every year," said Tom.

Collectors like to plant historic trees to mark a special occasion: a wedding, a graduation, a birth, or a death. "People have sentimental reasons for wanting these trees," Tom said. "They come to us because they want a tree that they have a connection to."

Universities have been in touch with the Hunters about planting a grove of literary trees connected to famous authors. "That's a cool idea, that the students can study an author and then come out and visit a tree that's associated with them," Tom said.

Not all historic trees come from specialty nurseries. In 2010, the Historic Gettysburg-Adams County Preservation Society offered a hundred seedlings from the honey locust that stood at the dedication of the Soldiers' National Cemetery, when Lincoln delivered the Gettysburg Address. These one-time offerings are easy to miss, but avid collectors watch out for them and place their orders early.

NAMESAKES OF NOTEWORTHY HISTORIC TREES

AUTHOR OF *ROOTS*

The Alex Haley Pecan, grown from seeds collected at the home of Haley's grandparents, which is now the Alex Haley Museum in Henning, Tennessee.

DISABILITY RIGHTS ACTIVIST AND AUTHOR

The Helen Keller Water Oak, grown from seeds of a tree she often wrote about, collected from the grounds of Ivy Green, which is now a historic site in Tuscumbia, Alabama.

PIONEERING AVIATOR

The Amelia Earhart Maple, collected from seed at the Amelia Earhart Birthplace Museum in Atchison, Kansas.

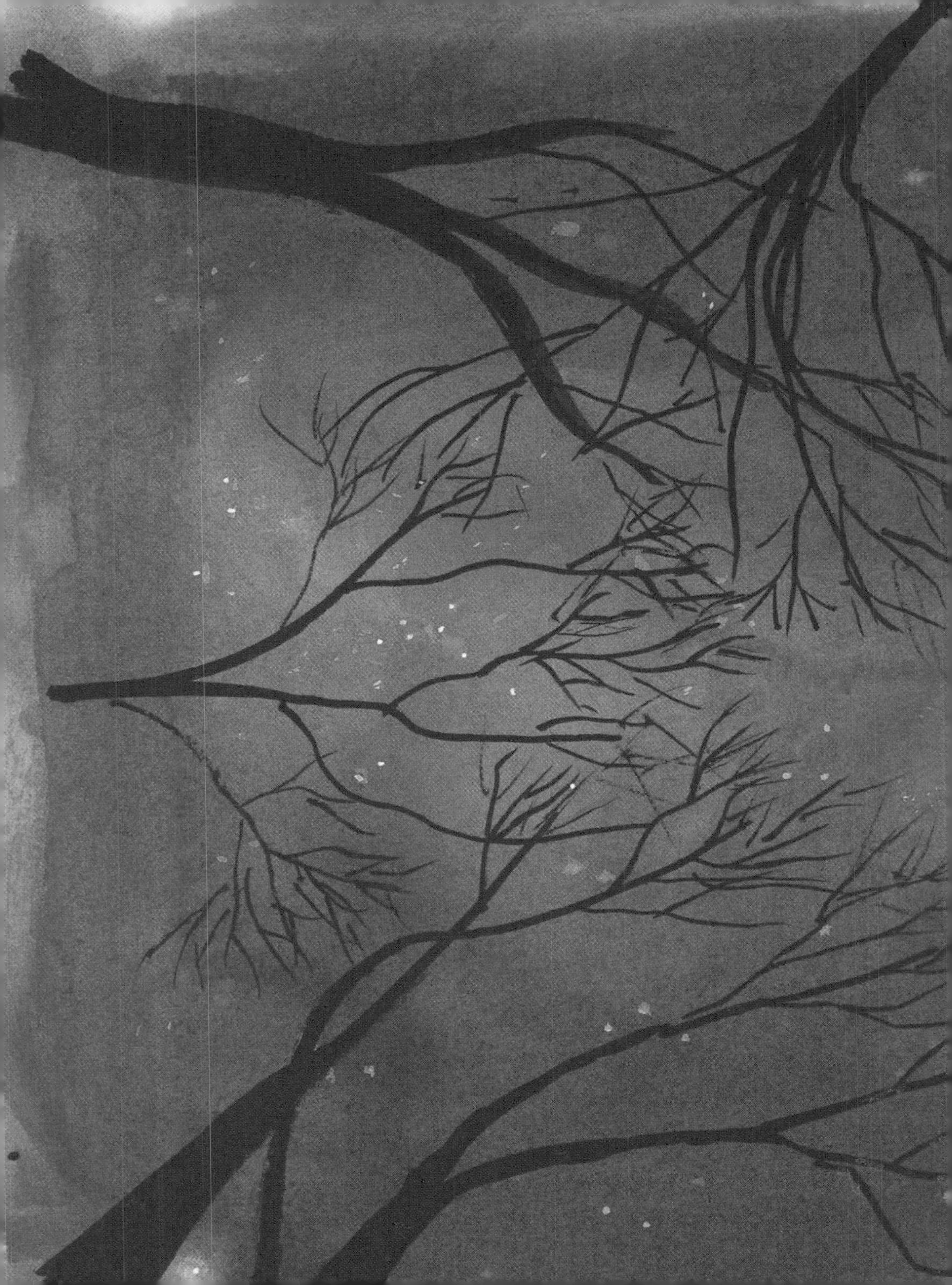

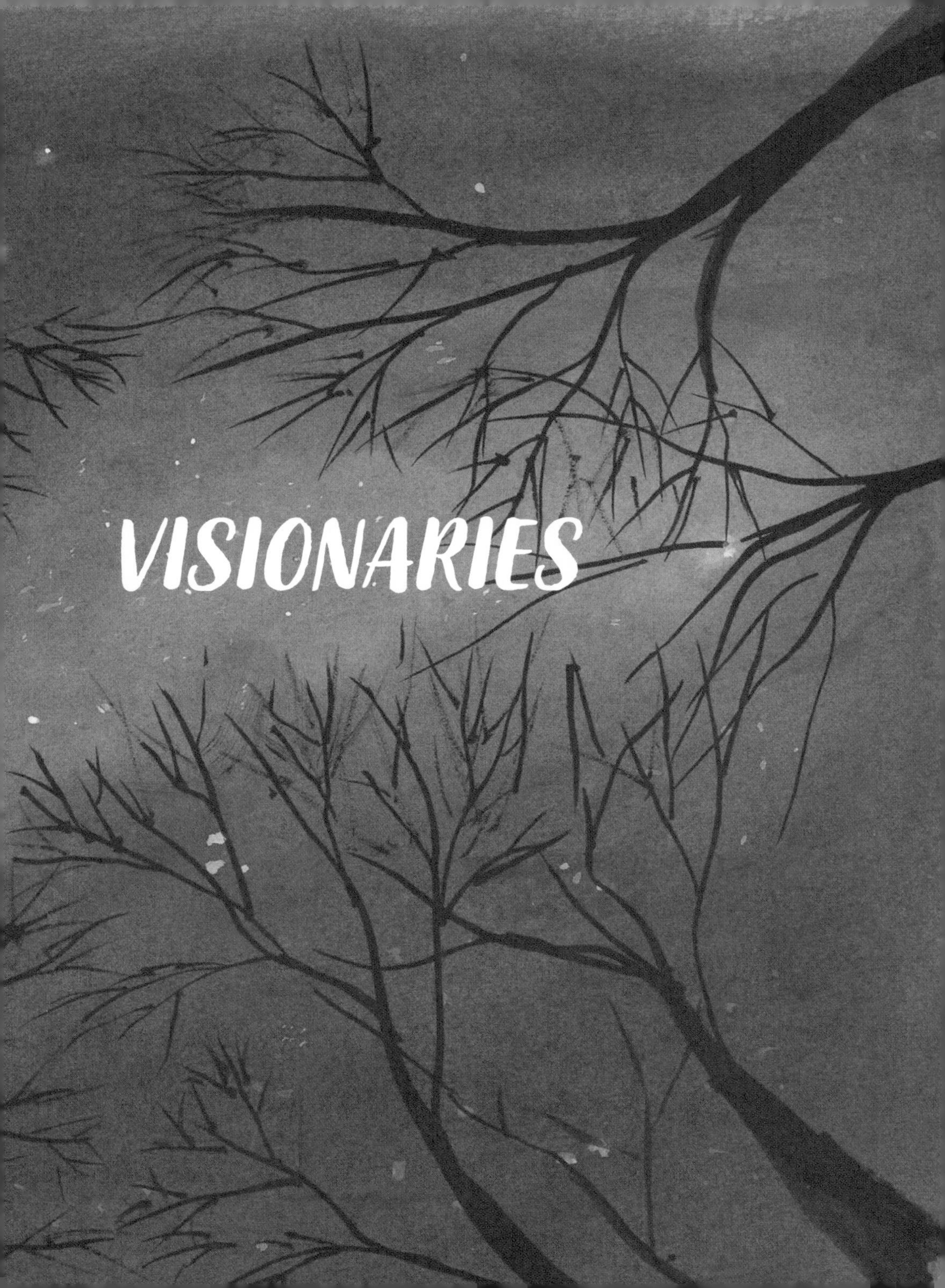

VISIONARIES

"I'VE WON MY THREE GOLD MEDALS...LET MARTY AND SAM RUN."

THE GOLD MEDALIST

JESSE OWENS
Cleveland, Ohio

WHEN JESSE OWENS STUNNED THE WORLD BY WINNING FOUR
gold medals at the 1936 Summer Olympics in Berlin, reporters described him "collecting" those four medals for his outstanding perfor-
mances in sprint and long-jump events. But he collected something else
in Berlin: four oak trees (*Quercus robur*), presented by the German
Olympic Committee to every gold medal winner that year.

At the time, these trees were called the Olympic Oaks, although
today they are also referred to as the Hitler Oaks. When Hitler's Nazi
Party took power in Germany in 1933, the International Olympic Committee had already designated Berlin the host city of the 1936 Summer
Olympics.

Organizers questioned that decision after Hitler's rise to power, and
calls to boycott the games began to surface around the world. Ultimately, the committee was persuaded by assurances from Germany that
Jewish athletes would be allowed to compete. In practice, however,
German sports clubs didn't admit Jews, making it impossible for them
to participate. Hitler's government made some efforts to conceal its
anti-Semitic agenda, such as temporarily removing signs that banned
Jewish people from public places. But ultimately, the 1936 Games provided a televised platform for the Nazi Party to project its image. It also
offered an opportunity for Black and Jewish athletes to refute the doc-
trine of Aryan superiority with their stellar achievements.

Jesse Owens was the most successful competitor at that year's games, and perhaps the most famous as well. His earlier achievements in track and field had already astonished Americans and dominated headlines. His accomplishments in Berlin were no less extraordinary.

He took gold medals in the 100-meter dash, the long jump, and the 200-meter sprint. But Owens's fourth gold medal was tainted by the anti-Semitism that marred the 1936 Games overall. Two Jewish athletes on the American team, Marty Glickman and Sam Stoller, were benched just before the trials for the 400-meter relay. The coach announced that they'd be replaced by Jesse Owens and Ralph Metcalf. Various excuses were given, but it appeared that the coaches were trying to appease Hitler. In a later interview, Glickman remembered that Owens said, "Coach, I've won my three gold medals . . . I'm tired. I've had it. Let Marty and Sam run, they deserve it." Glickman recalled the coach pointing at Owens and telling him, "You'll do as you're told."

Owens and Metcalf debated privately over what to do, but ultimately

they decided that it was better to compete than to step aside and let the Germans take the gold. Owens ran the relay and won his fourth gold medal. Glickman remembered it as the most humiliating episode of his life. He thought he'd be back in four years, but none of those men were able to compete again. Because of World War II, the next Olympics wouldn't be held until 1948.

Though 141 gold medals were awarded in 1936, most reports from the time state that only 130 oak saplings were handed out. Some were undoubtedly thrown away, either in disgust over the Nazi symbolism that relied on oak trees and forest images, or for ordinary lack of interest. Some died in hotel rooms or didn't survive long after they were planted. But a few of them are still alive around the world, almost ninety years later.

Owens managed to keep his alive. In a 1966 documentary he said, "And what about the oak trees that were given to me to plant? One I gave to the Rhodes High School in Cleveland, Ohio, the city where I spent my youth. One has flourished in the backyard of my mother's home in Cleveland. And one stands among the cherished mementos on All-American Row at Ohio State University, where I spent my college days. And the fourth one? The fourth one, unfortunately, has died."

Only the Rhodes High School tree survives today, and it's in poor health. Thanks to the efforts of community development groups and a local arboretum, cuttings were taken from it and successfully propagated. Those saplings have been planted at Rockefeller Park in Cleveland, at the new Jesse Owens Olympic Oak Plaza.

Jeff Verespej, the community organizer who spearheaded the effort, had no idea how difficult it would be to preserve this part of Owens's legacy. "I didn't know how hard oak trees were to graft," he said. "It's like a one in four success rate, maybe. But now we have ten or so that we're going to be able to plant. I hope people can learn something from these trees. I think every American needs to know this story better."

A SELECTED LIST OF SURVIVING OLYMPIC OAKS

 American relay runner **KENNETH CARPENTER** and discus winner **FOY DRAPE** planted their trees at the University of Southern California, where they were marked with plaques. One has died and been replaced with a tree grown from John Woodruff's acorns.

 Argentinian polo medalist **ROBERTO CAVANAGH** planted his tree near the headquarters of the Argentine Polo Association in Buenos Aires.

 Finnish poet **URHO KARHUMÄKI** won a gold medal in the mostly forgotten arts category, which once awarded medals to poets, painters, and sculptors. Its stump stands near his grave in Tervalampi, Finland.

 French weightlifter **LOUIS HOSTIN**'s tree can be found at the Parc de l'Europe in St. Étienne, France.

 American high-jump champion **CORNELIUS JOHNSON** planted his oak at his mother's house in Los Angeles, where it is the subject of a heated battle between a real estate developer and local preservationists.

 New Zealand track and field medalist **JACK LOVELOCK** planted his oak at Timaru Boys' High School, where it is still alive today. Acorns from this tree have been widely distributed and planted throughout New Zealand.

 Dutch swim team members **RIE MASTENBROEK, WILLY DEN OUDEN, TINI WAGNER, JOPIE SELBACH,** and **NIDA SENFF** are most likely responsible for the oak planted along the canal behind Amsterdam's Olympic Stadium.

 German cyclist **TONI MERKENS**'s oak can be found near the stadium in Köln, Germany.

 Swiss gymnast **GEORGES MIEZ** planted his near a stadium in Winterthur, Switzerland, where it thrives today.

 American track and field champion **JESSE OWENS**'s tree at Rhodes High School has survived and been grafted and propagated for new ceremonial plantings at Rockefeller Park in Cleveland, Ohio.

 American track and field medalist **JOHN WOODRUFF** planted his tree near the stadium in Connellsville, Pennsylvania, where acorns are regularly collected to cultivate another generation.

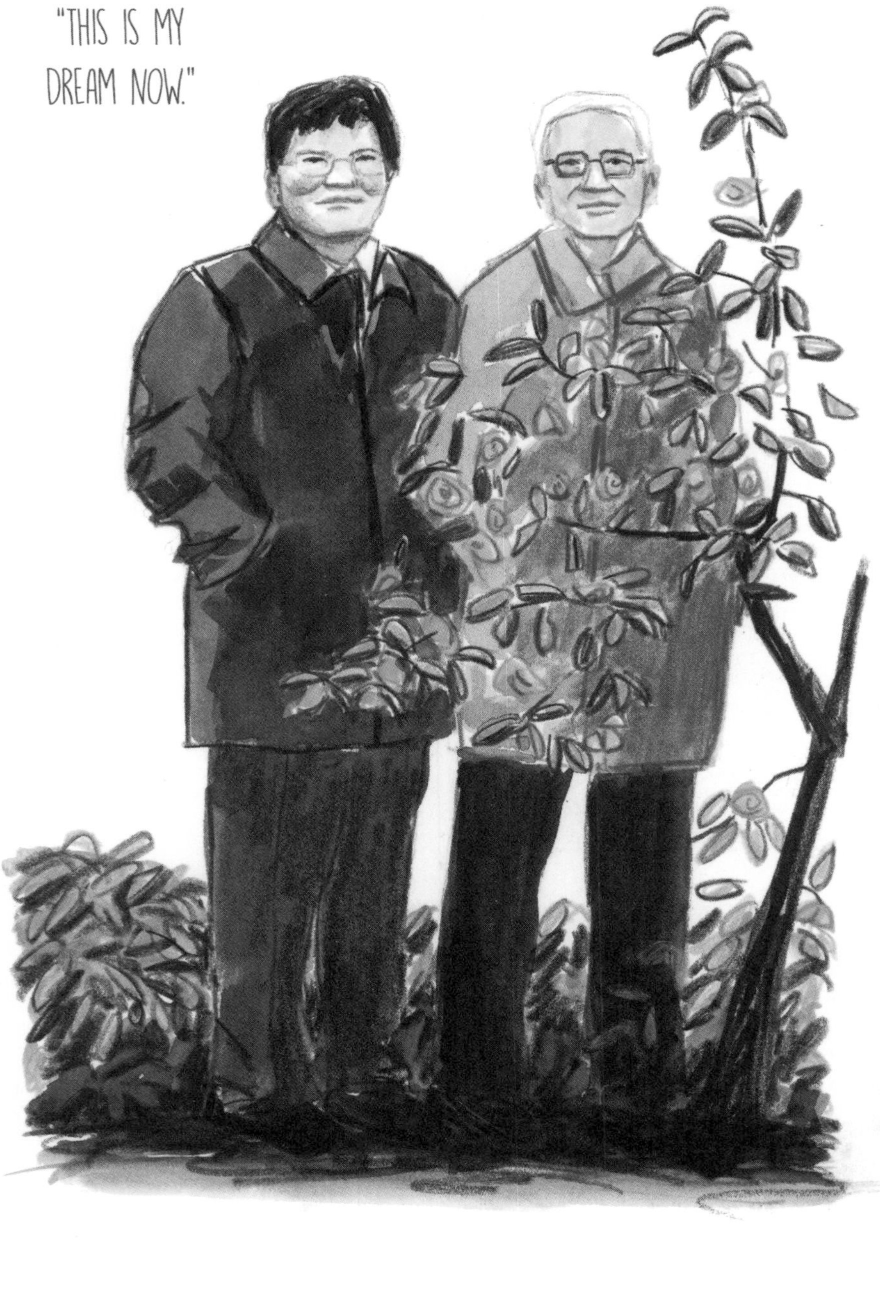
"THIS IS MY DREAM NOW."

THE EXPATRIATE
CARL FERRIS MILLER
South Korea

BOTANICAL HISTORY IS REPLETE WITH TALES OF BOTANISTS, mostly European and North American, who travel to faraway lands and bring back seeds and cuttings of exotic plants to be named, cultivated, and distributed at home. Very few of them choose to stay in those faraway countries, become citizens, and help to further the botanical work already underway.

Carl Ferris Miller is a notable exception. He wasn't a botanist—his passion for trees came later in life—but his commitment to his adopted country of South Korea has made a lasting impact on botanical science.

Miller was born in Pennsylvania in 1921. He studied Japanese in college, which led to a tour of duty with the navy during World War II. He went on to serve in Korea as an intelligence officer, worked for government aid organizations following the Korean War, and eventually took a job at the Bank of Korea. His contributions there were significant: he helped to modernize the Korean banking system after decades of Japanese occupation, winning him honors from the Korean government. He also joined the national Korean bridge team, and although he never married, he adopted three Korean sons.

In 1962, he took a trip to the fishing village of Chollipo, southwest of his home in Seoul. He was just after a little sun and swimming, but a villager persuaded him to buy a plot of vacant land. It sat idle until 1970, when he established a weekend home there to get away from Seoul's worsening air quality. Right away he decided that the property could use some trees.

This began a foray into tree collecting that would occupy the rest of his life. He scoured the countryside for interesting trees he could raise on his property. Villagers were eager to offer up parcels of unused land for sale, and soon he'd amassed about five hundred acres and filled it with his collection of hollies, magnolias, camellias, and maples, among others. In 1979, the Chollipo Arboretum became a nonprofit. That same year, Miller became a Korean citizen, one of the first Americans to do so, and took the name Min Pyong-gal.

Miller died in 2002, but the Chollipo Arboretum's director, Yong-Shik Kim (pictured on page 280 with Miller in 1981), was eager to explain the importance of his contributions to Korean botany. "Before he started the arboretum, the tree research in Korea was for forestry. For me, studying in the 1970s, I learned forestry and dendrology. Nothing about ornamentals."

At the time, he recalls, Korean botanists rarely worked with anyone outside their own country. "We Koreans were not open to outsiders and had very limited chances to communicate or exchange information. But Miller could speak English freely. He could communicate with international horticultural societies. I think that was a very important thing. He was very active and energetic. He awakened us to what was possible."

Miller didn't keep this knowledge to himself: he paid to send Korean botany students to study around the world. He helped them establish relationships with important institutions like Kew Gardens in London, Morris Arboretum in Philadelphia, and Longwood Gardens in Pennsylvania. "Korean society was very shy to work with others," Kim said. "But we had to collaborate and work together. He was the person to teach us this. It took a long time. Can you imagine when was the first time the staff at a Korean arboretum appeared at an international botanic garden congress? It was 1993. The Korean National Arboretum did not open until 1999."

Though he had no formal training in botany, Miller figured out the inner workings of the horticultural world. According to Kim, he was the first person working inside Korea to name and register Korean plants with international plant societies. "We never studied how to collect in the wild, how to grow what we collected, how to document, and how to register them with international organizations," Kim said. "Miller found out how to do that."

Today the Chollipo Arboretum is host to seventeen thousand species and subspecies from all over the world, including many species previously unknown to the international plant community. Its horticultural library, founded with Miller's own collection, totaling sixteen thousand volumes, is one of only a few in the country.

When Miller died in 2002, he left the entirety of his estate to the arboretum, allowing his legacy to live on through the work of botanists like Kim. "We have a duty to make better gardens and to work with others," Kim said. "This is my dream now."

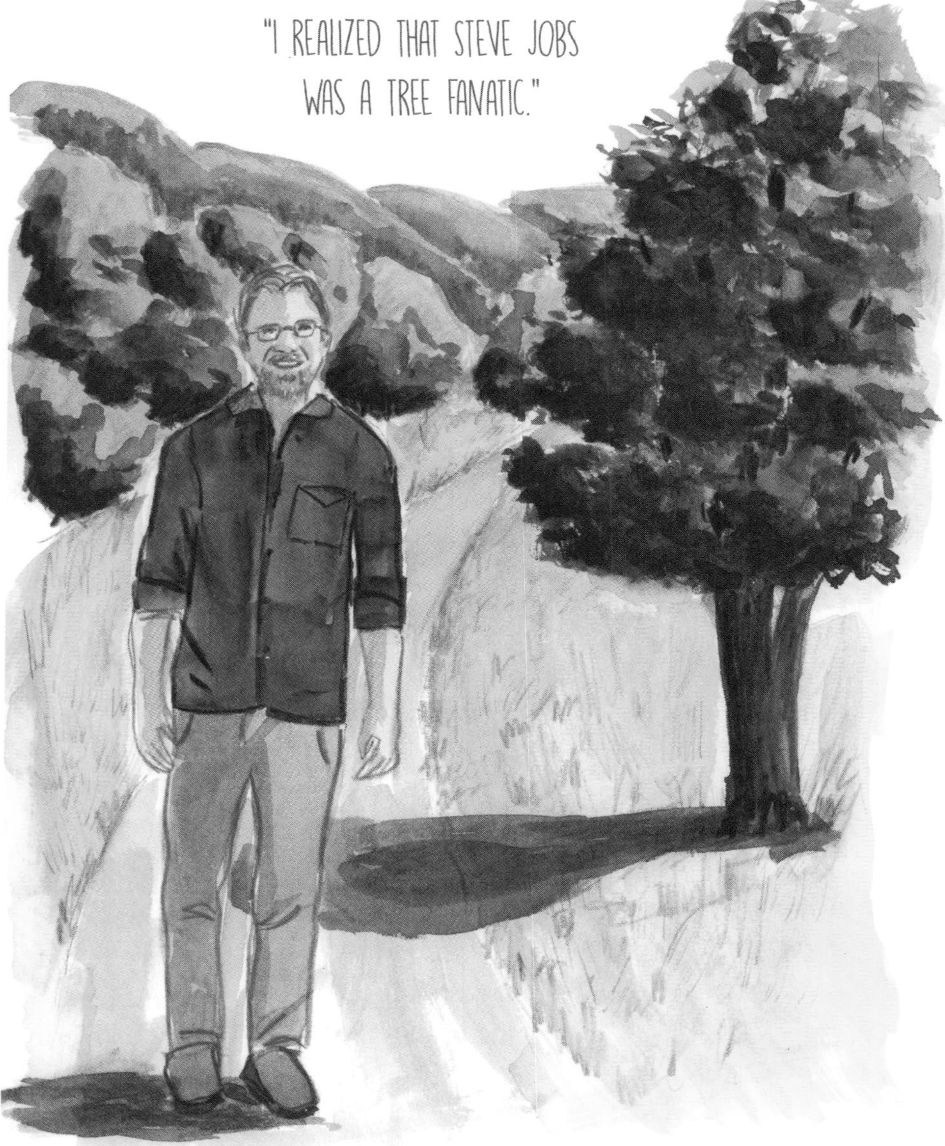
"I REALIZED THAT STEVE JOBS
WAS A TREE FANATIC."

STEVE JOBS'S ARBORIST

DAVE MUFFLY

Santa Barbara, California

As soon as Dave Muffly graduated from Stanford with a degree in mechanical engineering, he realized that he didn't actually want to be a mechanical engineer. "I went through a kind of post-college lost period," he said. He hung out with friends and got a job delivering pizzas. "After about a year of that, I confronted the reality that I needed a direction in my life, and pizza wasn't going to be it."

He felt called to do some kind of public service. At that point, in the late 1980s, he was hearing about climate change for the first time. He knew he was interested in ecological and social justice, but he wanted to do something physical. That led him to a nonprofit in Palo Alto called Magic.

Magic was founded in the 1970s as an intentional community dedicated to public service. The group owns three adjacent houses near the Stanford campus. The people who live there call themselves Magicians.

"It's a little like a monastery," Muffly said. "You get room and board, and you get taken care of. We lived kind of like ecological monks." The group focused on planting trees, both in urban areas and in open spaces around Palo Alto. One such space was the Stanford Dish Area, the site of a dish-shaped radio telescope jointly operated by the university and the government for military and space exploration purposes. Surrounding the telescope are miles of walking trails among gently rolling foothills. Magic was charged with planting oak trees in those hills.

This was exactly the sort of project Muffly was looking for. Like much of the land south of San Francisco, it had once been used for grazing, which had destroyed the native oak savanna ecosystem. With this project, he found his calling, and his community. "We were like smart urban hippies," he said.

When his stint at Magic ended, he continued to work on tree projects for various nonprofits, and he did some tree maintenance on the side, traveling mostly by bicycle. "A great way to learn trees is to bicycle. Cars are too fast, and walking is too slow. I've done thousands of urban bicycle miles with trees as my primary entertainment."

Over the years he returned several times to help manage Magic's oak reforestation project around the Dish. Once the first trees his group had planted were twenty years old, he noticed that some were struggling. The Magicians had failed to take into account the varying soil conditions in the area, which made an obvious difference in the health of the young trees. Now he had a chance to learn from those early mistakes.

Muffly didn't know it at the time, but while he was out planting trees, somebody was watching. "The Dish was Steve Jobs's favorite place to walk, and walking was one of his favorite things to do. He'd probably been walking around there for thirty years. He saw that whole project get planted."

One day in 2010, Muffly's phone rang. "Steve was looking to hire an arborist for Apple's new campus. I guess he told his headhunter to find the guy at Stanford."

Before long he found himself in a conference room with Steve Jobs. "The first thing I noticed was that it had windows that looked out onto oak-covered hills. And then I looked around at the paintings on the wall, and they were all paintings of California landscapes with oak trees. And I realized that Steve Jobs was a tree fanatic."

He described that moment as the biggest surprise of his life. "We really synchronized. He just got it on a fundamental level. When I said that we couldn't just plant native trees because we weren't going to have the same climate in thirty, forty, fifty years, he understood that right away. We had to diversify our tree portfolio in the face of an unknown, chaotic future."

The planned 175-acre campus, called Apple Park, would require Muffly to select and plant about nine thousand trees. Muffly spent years scouring nurseries across several states, and even walked abandoned Christmas tree farms, to find enough mature trees to fill out the landscaping. Fifteen acres were reserved for native grassland, and an orchard was planted to yield a succession of stone fruit and, of course, apples.

His role at Apple led to more tree-planting projects with corporate clients, universities, and private landowners. He tries to convince them to mix in non-native oaks, such as those from Mexico and the American Southwest, that he thinks might better withstand the climate of the future and serve as nodes of biodiversity in a crowded urban landscape. "I've had a long time to think about how to push back against the apocalypse," he said. "If you don't want to have a boring life, you might as well take on the biggest challenges there are out there. That's what I'm doing with these trees."

CORPORATE TREE COLLECTIONS

APPLE PARK
CUPERTINO, CALIFORNIA

Although the Apple campus's collection of nine thousand trees is not open to the public, a visitors' center offers rooftop views of the futuristic, flying-saucer-shaped office among the trees, along with a café and an Apple store.

BUFFALO TRACE DISTILLERY
FRANKFORT, KENTUCKY

The nation's oldest continuously operating distillery sits on four hundred acres, which also house an arboretum and botanical garden. Both garden tours and bourbon tastings are available year-round. The distillery's latest project, in partnership with the University of Kentucky, is an experimental plot of a thousand white oaks, the trees used to make bourbon barrels. They're studying genetic differences in oaks from across the country and experimenting with planting techniques to ensure the long-term sustainability of a tree that is of vital importance to bourbon drinkers and tree lovers alike.

THE DONALD M. KENDALL SCULPTURE GARDENS
PURCHASE, NEW YORK

PepsiCo's global headquarters sits inside an arboretum designed by renowned British landscape architect Russell Page. The public is invited to visit on a limited seasonal schedule, when they can tour an astonishing sculpture garden set among the trees featuring works by Alexander Calder, Louise Nevelson, Leonora Carrington, and Henry Moore, among others.

SALESFORCE PARK
SAN FRANCISCO, CALIFORNIA

Perched above the Salesforce Transit Center in downtown San Francisco is a remarkable small arboretum of more than six hundred trees gathered into collections representing the various regions of the world whose climates are comparable to the Bay Area. The park is open to the public year-round and features a number of rare and unusual trees, including a Wollemi pine, a weeping sequoia, and an 'Aptos Blue' coast redwood.

PACIFIC BONSAI MUSEUM
FEDERAL WAY, WASHINGTON

One of the world's finest bonsai collections was developed by the timber company Weyerhaeuser. The collection has since been placed under the care of a separate nonprofit, which operates it as one of only two bonsai museums in the United States. It's located just south of Seattle, next to the former site of Weyerhaeuser's corporate headquarters, which was once itself a 425-acre, tree-filled park. While the headquarters has been sold to developers, the bonsai museum remains, and sits next to a rhododendron garden.

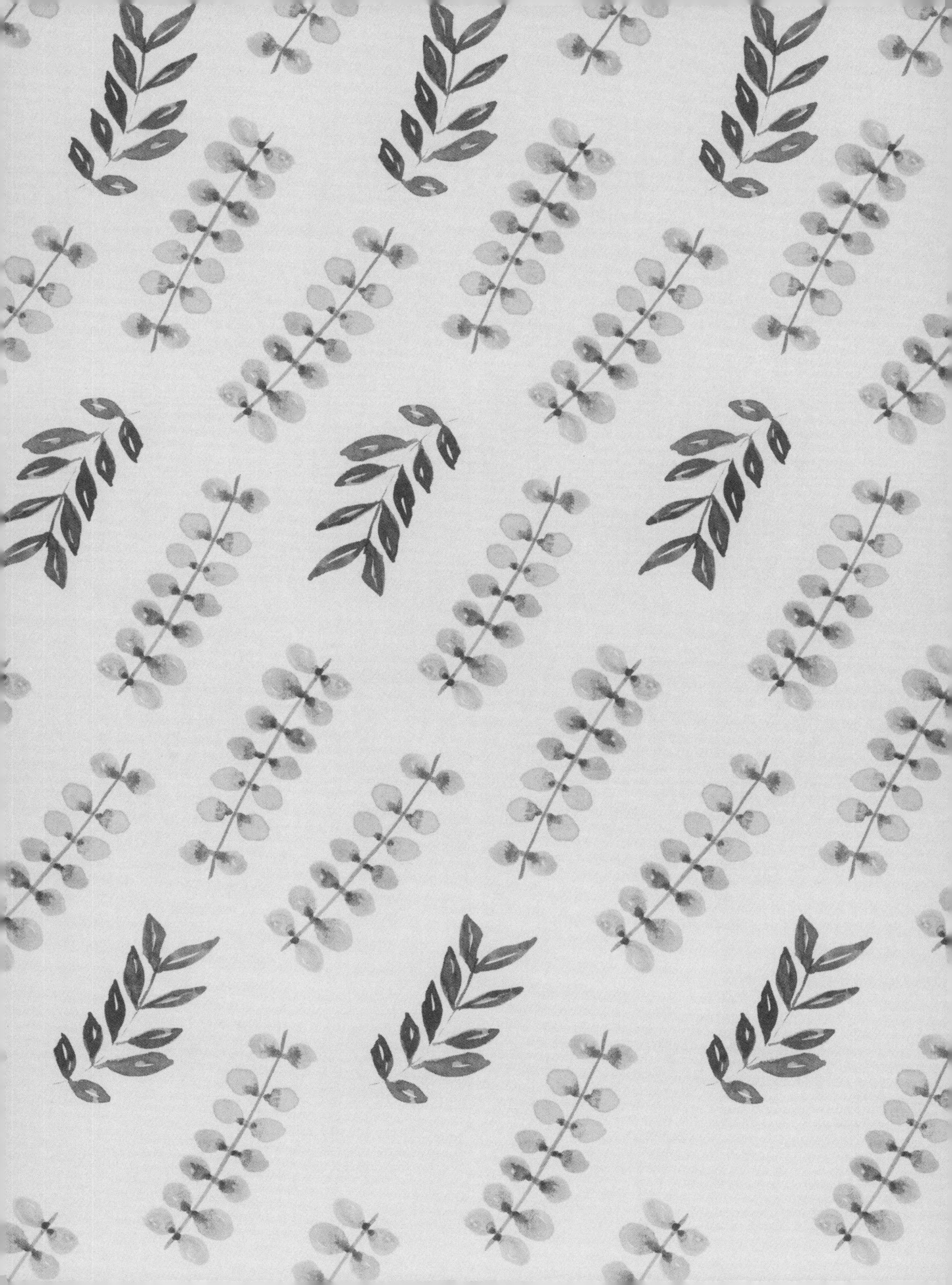

"A LOT OF PEOPLE WHO LOVE PALMS HAVE THAT FEELING THAT YOU'RE DEALING WITH OUR ELDERS."

THE POET

W. S. MERWIN
Haiku, Hawaii

"What's so interesting about William as a tree collector," said Sonnet Kekilia Coggins, director of the Merwin Conservancy, "is that he never would've thought of himself as a collector. There's something about ownership and control that's inherent in the idea of a collection. He liked to say that a person can't make a forest. Only a forest knows how to grow a forest."

New York–born William Stanley Merwin had already been awarded the first of two Pulitzer Prizes for Poetry and published many dazzling poetry collections when he traveled to Maui in 1976. He'd come to study with Zen teacher Robert Aitken. After knocking around in a series of makeshift rentals, he heard of a cabin for sale on three acres of land. The property had been entirely deforested by decades of destructive practices: trees cut down for firewood, land given over to cattle grazing, streams diverted to supply water to sugarcane plantations, and ill-conceived pineapple farming practices that wrecked the soil.

Nothing was left but overgrown grasses and the non-native, invasive Christmasberry tree, *Schinus terebinthifolius*. When Merwin saw the land, he recognized a long-cherished wish: "I had long dreamed of having a chance, one day, to try to restore a bit of the earth's surface that had been abused by human 'improvement.'" He bought the property and lived there for the rest of his life. When adjacent parcels were offered for sale, he bought those too, expanding the site to nineteen acres.

He hoped to return the land to a native Hawaiian rainforest, but the soil was too poor, and the land too exposed, to support those plants. The first trees he planted were non-native ironwoods, *Casuarina equisetifolia,* which would grow quickly but not invasively. Those trees created a canopy, and that canopy created the conditions for future plantings.

Over time he came to realize that it might never be possible to restore the land to an entirely native forest. The soil, he speculated, had been stripped of the microorganisms necessary to sustain native trees. Exotic pests and diseases were ever present, making it impossible to grow a tree like the native *Acacia koa*. He planted eight hundred of them, and they all died.

But he saw another possibility: he could grow as many species of palms from around the world as he could get his hands on. And once that desire possessed him, he did start to behave like a tree collector, even if he wouldn't have liked the label. He began exchanging letters with palm tree nurseries and horticulturists, swapping seeds, and importing rare and wonderful palms from around the world. He once described a feeling of kinship he and others had with these ancient trees—widespread sixty million years ago—this way: "A lot of people who love palms have that feeling that you're dealing with our elders."

Over the next three decades, he planted about 850 different species of palms totaling several thousand individual trees. For years he planted them at the rate of one a day. He never kept a list or made a map, and he might go years without visiting some of them. He didn't arrange the palms for display, but planted them where he hoped they would do best, often putting in several identical trees at once and allowing them to compete the way they would if they'd emerged as seedlings around a mother tree.

Like a collector, though, he had his prized specimens. An extremely rare palm, *Tahina spectabilis,* native to Madagascar, was entirely un-

known to botanists when it was found growing on a cashew plantation in 2007. Thanks to some quick work by botanists at Kew Gardens, seed was collected and distributed around the world so that it could be preserved. Dr. John Dransfield at Kew made a gift of one of these palms to Merwin, who was, by then, known as a respected and knowledgeable palm collector.

Of his role in conserving these trees, Merwin said that some of them only survived in cultivation "because of nuts like me. . . . If you leave it to the pros, they're better at it than we are, but there are fewer of them, so things are going to get lost."

Merwin established a conservancy before his death in 2019, ensuring that the palm forest would be preserved and that his house could serve as a retreat for writers and artists. His plant collection has now been

meticulously cataloged, the property is open to visitors on a limited basis, and residencies and fellowships support artists working across a range of disciplines.

Although Merwin is best known for his poetry, he felt his literary work was inextricably linked to his trees. He even liked to bury his mail at the base of trees, an act that any famous writer, overwhelmed with correspondence, unsolicited manuscripts, and invitations, would admire. His wife, Paula, said, "William's true life, that he truly loves, is writing poetry in the morning, and that involves reading and thinking and drinking tea and looking out at the palms, and then in the afternoon, potting and planting. Planting trees."

THE EXTRAORDINARY JOURNEY OF *TAHINA SPECTABILIS*

In 2005, Xavier Metz spotted some enormous, unfamiliar palms on the cashew plantation he managed. He shared photos with a friend, who posted the pictures to the discussion board of the International Palm Society. Kew botanist John Dransfield saw the pictures and asked a student from Madagascar, Mijoro Rakotoarinivo, to find the trees.

It quickly became clear that this was not just a new species, but an entirely new genus. It was given the name *Tahina,* after one of Metz's daughters. Botanists would eventually learn that the tree flowers abundantly just once, at the end of its life. Fortunately, one specimen was in bloom, and there was plenty of seed to collect.

The botanical community came together to organize a plan. Seed would be collected and distributed to carefully chosen locations to preserve the trees. Additional seed could be sold, with proceeds benefiting the local community.

Locals worked hard to protect the trees from the ravages of wildfires, animal predators, and meddlesome foreigners. The income helped renovate a school, build a well, and fund other development projects.

Today the palm can be seen at public gardens around the world, and John Dransfield's gift is still thriving at the Merwin Conservancy.

ACKNOWLEDGMENTS

Thanks go first to every tree collector who shared their story with me and introduced me to other collectors. Their names appear in these pages, and I hope I have done justice to their stories.

I'm also grateful to the many collectors, botanists, and arborists who offered their insight and expertise, including Francisco Arjona, Jennie Ashmore, Sherry Austin, David Benscoter, Augustin Coello, Dave Dexter, Fred Durr, Bill and Heather Funk, Jose Miguel Gallego, Eliza Greenman, Vanessa Handley, Gabriel Hemery, Jill Hubley, Phyllis and Tom Hunter, Vanessa Labrouse, Peter Laharrague, Louis Meisel, Jeff Meyer, Sue Paist, Matt Ritter, Philippe de Spoelberch, Rose Tileston, Simon Toomer, John Uri, Robert Van Pelt, Dennis Walsh, Richard Weir, Jeff Winget, and Paul Wood.

I'm also very grateful to the translators and interpreters who made it possible for me to speak to people from around the world. They took on this project as their own and added their valuable questions and observations to our conversations. Thanks to Malgorzata Bruzek, Claudia Dornelles, Tatiana A. de Faria, Van Luu Thi Hong, Vivek Kumar, Yukari Yagura, and Hani Yeo.

Thanks as always to my agent, Michelle Tessler, and to Jessica Dacher for her help with the proposal. Endless gratitude to Hilary Redmon, Robin Desser, and everyone at Random House who brought this book to life.

Finally, many thanks to my husband, the book collector who loved the idea of this book from the beginning, P. Scott Brown.

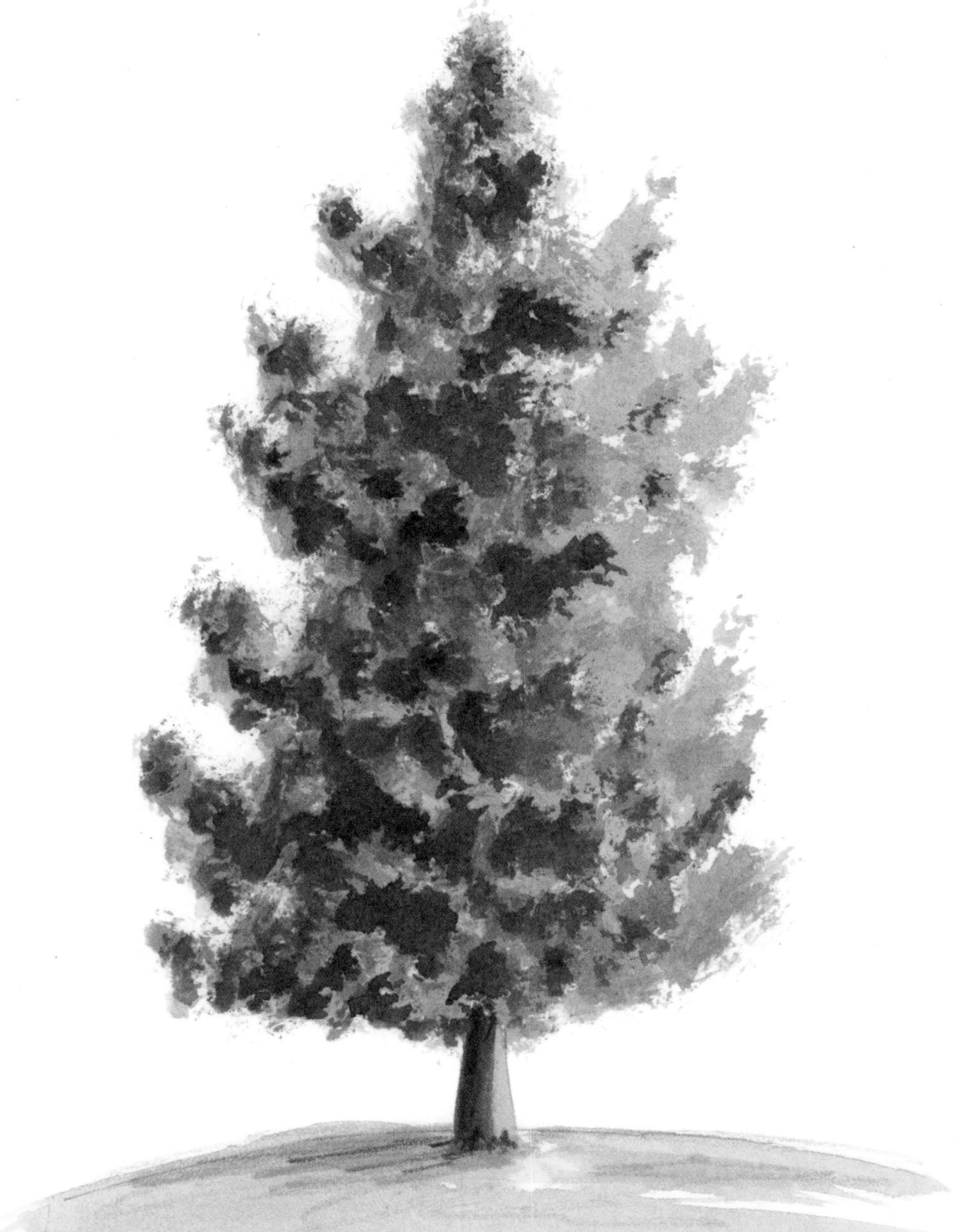

NOTES

The quotes in this book come mostly from personal interviews with the subjects. Sources of quotes from published materials are provided below, along with suggestions for further reading.

PRIVATE TREE COLLECTIONS YOU CAN VISIT

96 **"Start looking into oaks"** Mark Griffiths, "The Life of the Oak Tree Collector: 'You See There's Only One Sensible Course of Action: Collect the Lot,'" *Country Life*, August 31, 2019.

JOEY SANTORE

145 **"city kind of dropped the ball"** quoted from Santore's YouTube video, *Tony Santoro's Guide to Illegal Tree-Planting*.

EDWARD EVERETT HORTON

Material in this chapter comes from my review of Horton's archives at the University of Wyoming's American Heritage Center in Laramie, and quotes were cited from the following newspapers:

209 **"I was sitting up with a sick tree"** *Arizona Republic* (Phoenix, AZ), August 1, 1937.

209 **"Take my fan mail to prove it"** *Richmond Times-Dispatch* (Richmond, VA), June 13, 1937.

210 **"A bootlegger who was en route"** *Democrat and Chronicle* (Rochester, NY), December 21, 1959.

211 **"These things are so unpredictable"** *The Post-Star* (Glens Falls, NY), July 3, 1952.

HALL OF FAME: CELEBRITY TREE COLLECTORS

212 **"My life now is just trees"** *Judi Dench: My Passion for Trees,* BBC, 2017.

YNÉS ENRIQUETTA JULIETTA MEXÍA

For more on Mexía, read Durlynn Anema's *The Perfect Specimen: The 20th Century Renown Botanist—Ynes Mexia* (National Writers Press, 2019).

219 **"bear with the marriage relation"** Mexía to Petsito (Augustin Reygadas), July 31, 1911, Ynés Mexía papers, 1872–1963, BANC MSS 68/130 m, Bancroft Library, University of California, Berkeley.

220 **"Everyone tells me of all sorts"** Mexía, letter addressed "Dear Miss," October 24, 1926, Ynés Mexía papers, 1872–1963, BANC MSS 68/130 m, Bancroft Library, University of California, Berkeley.

222 **"She didn't care for people"** Nina Floy Bracelin, interviewed by Annetta Carter (oral history transcript), 1965, 1967; Ynés Mexía Botanical Collections, Bancroft Library, University of California, Berkeley, 1982.

222 **"Then we started on the long journey"** Ynés Mexía, "Camping on the Equator," *Sierra Club Bulletin* 22, no. 1 (February 1937): 85–91.

PLANT EXPLORERS AND THE TREES NAMED FOR THEM

For more on Joseph Banks, read the entertaining new biography *The Multifarious Mr. Banks: From Botany Bay to Kew, the Natural Historian Who Shaped the World* by Toby Musgrave (Yale University Press, 2020).

The story of David Douglas is told in *The Collector: David Douglas and the Natural History of the Northwest* by Jack Nisbet (Sasquatch Books, 2010).

Andrea Wulf's *The Invention of Nature: Alexander von Humboldt's New World* (Vintage, 2016) is a wonderful introduction to the life of von Humboldt.

Manuel Incra Mamani's story is told in Mark Honigsbaum's *The Fever Trail: In Search of the Cure for Malaria* (Farrar, Straus and Giroux, 2002).

For a complete biography of Frank Meyer, see *Frank N. Meyer: Plant Hunter in Asia* by Isabel Shipley Cunningham (Iowa State University Press, 1984).

A collection of Marianne North's own writings can be found in *Abundant Beauty: The Adventurous Travels of Marianne North, Botanical Artist,* with an introduction by Laura Ponsonby (Greystone Books, 2011).

SAVING JAPAN'S CHERRY TREES

For more on the remarkable story of the effort to preserve heritage cherry tree varieties in Japan, read *The Sakura Obsession: The Incredible Story of the Plant Hunter Who Saved Japan's Cherry Blossoms* by Naoko Abe (Knopf, 2019).

More on David Fairchild's role in bringing Japan's cherry trees to America can be found in *The Food Explorer: The True Adventures of the Globe-Trotting Botanist Who Transformed What America Eats* by Daniel Stone (Dutton, 2018).

260 **"There were once three hundred varieties"** David Fairchild, "The Ornamental Value of Cherry Blossom Trees," *Art and Progress* 2, no. 8 (1911): 225–226.

Eliza Scidmore's adventurous life is recounted in Diana Parsell's book *Eliza Scidmore: The Trailblazing Journalist Behind Washington's Cherry Trees* (Oxford University Press, 2023).

JESSE OWENS

276 **"Coach, I've won my three gold medals"** Marty Glickman's recollections appear in an interview published by the United States Holocaust Memorial Museum for an exhibit called "The Nazi Olympics, Berlin 1936." The recording is available on the museum's website.

277 **"And what about the oak trees"** Jesse Owens's account of the fate of his Olympic Oaks is quoted from the 1966 documentary *Jesse Owens Returns to Berlin*.

For more on Owens and the Berlin Olympics, read *Triumph: The Untold Story of Jesse Owens and Hitler's Olympics* by Jeremy Schaap (Mariner Books, 2015).

W. S. MERWIN

293 **"I had long dreamed"** W. S. Merwin, "The House and Garden: The Emergence of a Dream," *Kenyon Review* 32, no. 4 (Fall 2010): 10–24.

294 **"A lot of people who love palms"** Stefan C. Schaefer, director, *Even Though the Whole World Is Burning,* documentary, 2014.

295 **"because of nuts like me"** Schaefer.

296 **"William's true life"** Schaefer.

ABOUT THE AUTHOR

AMY STEWART is the *New York Times* bestselling author of *The Drunken Botanist, Wicked Plants,* and several other popular nonfiction titles about the natural world. She's also written seven novels in her beloved Kopp Sisters series, which are based on the true story of one of America's first female deputy sheriffs. She lives in Portland, Oregon.

amystewart.com
Instagram: @amystewart